D0860963

WEATHER ANALYSIS & FORECASTING HANDBOOK

TIM VASQUEZ

WEATHER GRAPHICS
TECHNOLOGIES

First edition
April 2011

ISBN 978-0-9832533-0-3

CS4LS E3

Printed in the United States of America

Weather Graphics Technologies
P.O. Box 450211 Garland, TX 75045
(800) 840-6280 fax (206) 279-3282
Web site: www.weathergraphics.com
support@weathergraphics.com

Contents

1 PHYSICS / 1
1.1. Mathematics / 1
1.2. Time / 2
1.3. Distance, direction, and velocity / 2
1.4. Mass, force, and pressure / 3
1.5. Temperature / 4
1.6. Density / 5
1.7. Water / 5
1.8. Coriolis force / 8
1.9. Wind forces / 8
1.10. Geostrophic wind / 9
1.11. Gradient wind / 10
1.12. Cyclostrophic wind / 11
1.13. Vorticity / 11
1.14. Horizontal coordinate systems / 13
1.15. Vertical coordinate systems / 13
1.16. Scale / 15
1.17. Atmospheric structure / 15
1.18. Global circulation / 17

2 OBSERVATION / 20
2.1. Observation networks / 21
2.2. Observation coding formats / 24
2.3. Temperature / 25
2.4. Dewpoint / 25
2.5. Wind / 25
2.6. Pressure / 26
2.7. Visibility / 27
2.8. Weather / 27
2.9. Clouds / 28
2.10. Upper air systems / 30

3 THERMODYNAMICS / 34
3.1. Phases of matter / 35
3.2. Adiabatic changes / 36
3.3. Stability / 38
3.4. Instability / 38
3.5. Soundings / 39
3.6. Sounding interpretation / 43
3.7. Instability quantification / 47
3.8. Potential instability / 48
3.9. Symmetric instability / 48

4 UPPER AIR ANALYSIS / 52
4.1. Constant pressure charts / 54
4.2. Long waves / 54
4.3. Short waves / 58
4.4. Divergence and convergence / 59
4.5. Vertical motion / 60
4.6. Jets / 62
4.7. Jet streaks / 66
4.8. Thermal advection / 68
4.9. Thickness / 69
4.10. Frontogenesis and frontolysis / 71
4.11. Vorticity / 73
4.12. Q vectors / 78
4.13. Isentropic analysis / 79

5 SURFACE ANALYSIS / 84
5.1. The surface chart / 85
5.2. Air masses / 87
5.3. Frontal concepts / 90
5.4. Cold front / 93
5.5. Warm front / 93
5.6. Quasistationary front / 94
5.7. Occluded front / 95
5.8. Dryline / 96
5.9. Outflow boundaries / 98
5.10. Sea/land breeze fronts / 99

6 WEATHER SYSTEMS / 102
6.1. Baroclinic lows / 103
6.2. Baroclinic high / 108
6.3. Cold-core barotropic low / 110
6.4. Warm-core barotropic low / 112
6.5. Cold-core barotropic high / 114
6.6. Warm-core barotropic high / 115
6.7. Arctic air outbreaks / 116

7 SATELLITE / 126
7.1. Satellite orbits / 127
7.2. Imagery types / 128
7.3. Satellite imagery limitations / 130
7.4. Clouds / 132
7.5. Patterns / 134

8 RADAR / 142
8.1. How radar works / 143
8.2. Reflectivity / 143
8.3. Velocity / 144
8.4. Spectrum width / 146
8.5. Dual-polarization data / 146
8.6. Problems and pitfalls / 149
8.7. Severe weather signatures / 151
8.8. VAD/VWP wind data / 153

9 CONVECTIVE WEATHER / 156
9.1. Thunderstorm structure / 157
9.2. Multicellular storms / 158
9.3. Supercells / 159
9.4. Mesoscale convective systems / 162
9.5. Wind profiles / 165
9.6. Tropical weather circulations / 169
9.7. Tropical cyclones / 174
9.8. Tropical cyclone forecasting / 178

10 PROGNOSIS / 182
10.1. The forecast process / 183
10.2. Numerical model concepts / 188
10.3. Numerical forecast production / 190
10.4. An overview of available models / 192
10.5. Limitations of models / 193
10.6. Climatological patterns / 195

Introduction

The vast majority of weather books fall into one of three categories. The "retail bookstore" category is made up of titles for novices that have lots of color pictures, plenty of meteorology history, and even a bit about how forecasting is done but not how to actually forecast. The "technical books" and journals are aimed at professionals, and they make up the core of meteorological knowledge. However they usually deal with only one specialized aspect of forecasting, concentrating on the underlying physics using algebra and partial differential equations and leaving it up to the forecaster how to apply the knowledge. The "college bookstore" titles are written to ground students in that same body of scientific method, and likewise few of these titles go forward with how to apply the information to the forecast process.

This book is a little different. I have always been interested in providing a book that goes straight from the basics into the deep material, allowing new forecasters hit the ground running and help prospective forecasters understand the kinds of problems that they'll soon encounter. Overall, my aim in writing this book, and its predecessor *Weather Forecasting Handbook*, has been to present a readable, comprehensive, yet sufficiently technical introduction for amateurs with a little bit of fun and humor thrown in. *It is not* designed to be a dry, exacting textbook or a scientifically-groundbreaking work. Therefore there isn't a focus on references and there is no effort made to analyze each concept in detail. What does appear is trivia, tips, anecdotes, history, and even a favorite recipe.

If an error or typo is discovered in this text, I ask as a courtesy that you inform me at servicedesk@weathergraphics.com so that it can be added into our errata and corrected in future editions. All corrections and suggestions are definitely acted upon rather than filed away in a drawer. The responsibility of making a book truly useful does not stop once it's in print. Visit this book's official site at www.weathergraphics.com for downloadable supplements, errata, and news about updates.

Readers of *Weather Forecasting Handbook* will notice that some of the content is the same. This book was intended to be something of a hybrid between a significant revision and an all-new book. I felt that the original handbook was too introductory and it was time to expand it further into the practice of operational forecasting. Rather than scrap the old version and its time-tested content, some of the material has been left in place or built upon.

Tim Vasquez
April 7, 2011 / Norman, Oklahoma
servicedesk@weathergraphics.com

Special notes

Brevity
This is an operational forecasting book title. In the interests of readability and brevity, the technical content, math, and references have been kept to a bare minimum. This is not to discredit the body of meteorology research that exists; certainly only a few of the core techniques are my own. The lack of references allows a lot of dry, technical material to be presented in the most readable, engaging style possible.

Antonyms
To eliminate extra text where it would prove distracting, meteorological conditions involving multiple states are sometimes written with a second set of conditions and outcomes in brackets. An example is the phrase *"if Q vectors point toward warm [cold] air, frontogenesis [frontolysis] is indicated"*. It is intended to be read two ways: one using only the non-bracketed nouns and adjectives, and another using the bracketed set instead. The above example states that warm air is associated with frontogenesis and cold air with frontolysis.

Signs in opposite hemispheres
Some passages in this book are written specifically for the Northern Hemisphere and are marked with an asterisk to indicate this fact. They can be reversed for use in the Southern Hemisphere; e.g. left rear is changed to right rear, cyclonic to anticyclonic, and south to north.

Examples
Many of the examples are constructed from NCEP DIFAX charts since they are very high resolution and designed specifically for monochrome print. Charts from the Internet are usually of resolution too low to print in book form, and the colors do not reproduce well in monochrome ink. However any Internet chart with the same information can be used the same way as the charts presented in this book.

Experience is not what happens to a man; it is what a man does with what happens to him.

-- *Aldous Huxley, author*
"Texts and Pretexts", *1932*

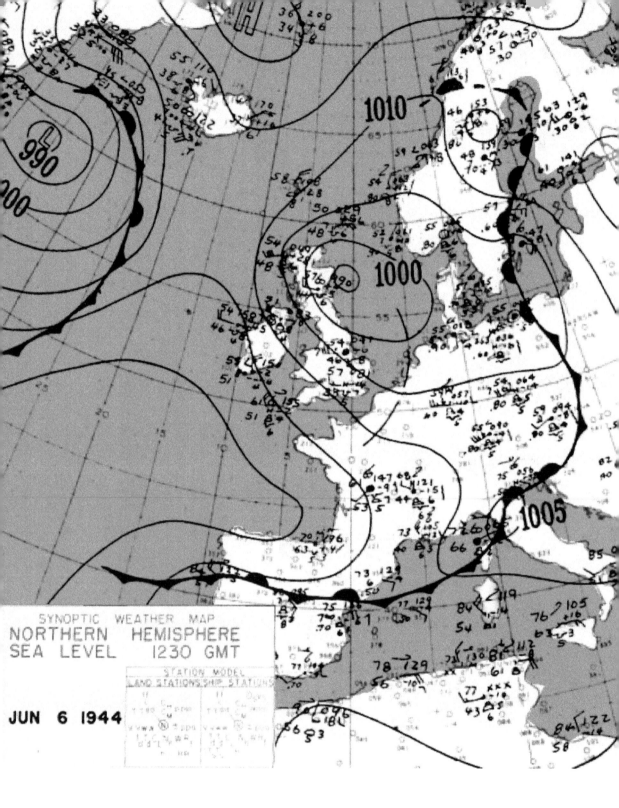

SYNOPTIC WEATHER MAP
NORTHERN HEMISPHERE
SEA LEVEL 1230 GMT

STATION MODEL
LAND STATIONS SHIP STATIONS

JUN 6 1944

1 PHYSICS

Avoid skipping this chapter! It is the bedrock of the material that will follow throughout this book and will make it much more understandable. Every effort has been made to avoid souring the exciting, enjoyable process of weather forecasting with bits and pieces that have nothing to do with the actual forecast process. However take into consideration that a business traveller enroute to Madrid to meet a client would likely pack an iPod with some sort of Spanish audio tutorial. Likewise, this chapter serves as a valuable crash course that conveys not only the language but many important conventions and customs. Only the basics are contained here.

1.1. Mathematics

It's not possible to completely eliminate math from a proper introduction into forecasting, because meteorology describes a *system* made up of closely-linked and interdependent properties. The forces that act on an air parcel (an imaginary cube) are firmly tied to many atmospheric properties, and a simple math equation is often much more succinct than a wordy statement.

Consequently, a very basic amount of math must be included to demonstrate in clear terms how and why a simple change affects other properties. The result could very well be significant weather. Fortunately, the equations presented here *are not meant to be solved*, but rather simply to make it clear how one variable affects another without a wordy explanation. This also paves the way for understanding more complex equations in other forecasting books. All the reader needs to do to understand the equation is to consider the effect each variable has on the property being solved.

For instance, consider the equation $D=rv$, where D is the distance a car will go, r is the mileage of a car (perhaps miles per gallon) and v is how much fuel we put in this car. Since we must multiply r and v to find D, it follows that increasing either r or v will cause D to increase; in other words, a car with better mileage or more fuel will achieve this.

We can also create a thought experiment and hold D constant. To do this, any increase in r will need to be offset by a decrease in v. To phrase this in English terms, a car with lots of fuel and poor mileage will go about the same distance as a car with a little fuel and good mileage.

We can re-arrange the above equation to get a division equation: $r=D/v$. This tells us that the car's mileage is obtained from the distance travelled (the numerator) divided by fuel used (the denominator). Increasing the numerator value increases the

Watch those units
The Mars Polar Orbiter, which was launched in December 1998, was destroyed when it unexpectedly plunged into the Martian atmosphere and crashed. The cause? A contractor, Lockheed Martin, provided a table of navigation data in units of "pounds per second" to the Jet Propulsion Laboratory (JPL). They were expecting "grams per second".

Title image
Surface chart on June 6, 1944, the date on which a monumental Allied ground invasion began in continental Europe. Poor weather had affected the English Channel region, but by this date the northern French shores were on the back end of a frontal system well within an area of cold air advection. The ridge to the west ushered in a period of fair weather. *(NOAA)*

result, i.e. if a greater distance is travelled, the mileage is better. But increasing the denominator decreases the result, suggesting that if more fuel was consumed, the mileage was lower.

1.2. Time

One of the most familiar units of measurement is the second. This is given by the abbreviation s and is the basis for expressions of velocity and acceleration in the sections ahead.

Weather systems and centralized products regularly flow across time zones, so to help reduce confusion and simplify the workload, meteorologists preferentially use Universal Coordinated Time (UTC), also known as Greenwich Mean Time (GMT) or Zulu time (Z). In simple terms, UTC is the time and date in London, England, ignoring any daylight saving time ("summer time") rules that are observed there. Forecasters should be thoroughly familiar with how to convert their local time zone to UTC and back. In the United States, Eastern Standard Time is converted to UTC by adding 5 hours, or 4 hours when Eastern Daylight Time is in effect.

1.3. Distance, direction, and velocity

Distance is the amount of space separating two points. It is expressed in meters (m), though since meteorologists deal with vast distances, the expression kilometers (km) is often used. Two-dimensional extent is area, a. The SI system prescribes it is given in square meters, m^2. Three-dimensional extent is volume, v, given in m^3. These units often come up in radar meteorology.

Velocity, given by the abbreviation v, is used to describe distance covered in a specific time. It is given in $m\ s^{-1}$ (i.e., m/s, meters per second). In non-physics use, meteorologists often express velocity as knots (kt), one of which equals $0.514\ m\ s^{-1}$ or 1.15 mile per hour. It should be noted that the abbreviation v is sometimes used to express volume and distance, so some caution is warranted when it appears in equations.

A velocity magnitude that changes with respect to time is acceleration, $m\ s^{-2}$ (i.e., m/s/s, or meters per second of change per second). Earth's gravity imposes an acceleration of about $9.8\ m\ s^{-2}$, which means that in the first second, an object unaffected by air resistance is travelling at $9.8\ m\ s^{-1}$ after the first second and $19.6\ m\ s^{-1}$ after the next second.

Velocity has not only a magnitude, or speed, but also a direction. This is expressed in degrees relative to true north, the direction of the North Pole. Meteorologists must be extremely attentive to whether a direction describes where wind is *blowing from* or *blowing to*. Traditionally, the phrase *wind direction* expresses which direction a wind has originated from; i.e. a paper bag blowing eastward is being blown by a westerly wind. However, *wind vector* expresses which direction winds are blowing toward; the wind blowing this paper bag has an eastward vector. In operational meteorology, wind vectors are used primarily when working out physical equations and when using the hodograph, while nearly all weather charts use wind direction instead.

Fortunately, when it comes to actual weather charts, the use of wind symbols in meteorology is never ambiguous. Vector symbols, lines with arrows on the tip, are always reserved for wind vectors. The arrow points downstream. Meanwhile, the so-called "wind barb", a shaft which contains barbs and pennants to indicate wind speed, is always an indicator of wind direction and the shaft points upstream.

Meteorologists are often interested in how the *wind direction* changes with respect to time, or across a given space. If the direction becomes more counterclockwise in some respect, this is referred to as *backing*. If the direction becomes more clockwise, this is referred to as *veering*. This is the definition which applies in the Northern Hemisphere. The situation is very ambiguous in the Southern Hemisphere, because the meaning of backing or veering differs according to whether international convention or American convention is used. Context and common sense must be used to determine the meaning of the words in such instances.

1.4. Mass, force, and pressure

One depiction of mass in popular culture was in a 1986 comedy skit on *Saturday Night Live*, in which the late comedian Phil Hartman played a science presenter who knocked a 1400-pound chunk of osmium metal off its pedestal, which was then depicted to crash endlessly through all the floors below. Ostensibly this item had a mass of 600 kg, but what do we mean by mass? Does its weight on a bathroom scale or its tendency to fall through the floor with great force give it the qualities of mass?

If this imaginary osmium was taken into outer space where gravity did not exist, it would be perfectly harmless if its pedestal collapsed, and it would weigh zero pounds on a bathroom scale. However, it would still have a mass of 100 kg and would certainly be a deadly weapon if thrown across the spaceship. Mass actually is a measure of the total amount of atomic matter a sample contains. It also is an expression of the types of energy that bind it together, and its inertia, or resistance to being moved.

The Laws of Thermodynamics
These laws are the building blocks for learning about energy exchange between a system and its surroundings. Their understanding is not critical for the operational forecaster, but they form the bedrock for understanding how heat is exchanged.

First Law: Energy cannot be created or destroyed; it can only move or change form. A campfire does not create energy; it liberates stored energy. Likewise energy is not lost when the campfire goes out; it simply moves elsewhere or changes form.

Second Law: In a closed system thermal energy will only flow from a hot spot to a cold spot, a process called heat transfer. If any work is done (activity which moves mass), thermal energy is transferred through out the system. Once the system's thermal energy is fully uniform, no more heat transfer can occur and work cannot be done. This process is called entropy.

Third Law: The entropy process is only fully reversible at absolute zero. However, no system can achieve absolute zero.

In summary: Since thermal energy cannot be created and must be taken from an active or stored source, it's often said "you can't win". Since waste heat cannot be completely recovered, it's said that "the best you can do is break even". Since expended energy can only be fully recovered at absolute zero, which is unattainable, it's said "you can only break even at absolute zero". It's theorized that the universe will eventually reach maximum entropy, after about 10^{100} years where no energy is available for work.

Mass is expressed with the symbol *m* and is given in grams (g) or kilograms (kg).

When acted upon by acceleration, mass gains a new property: force, written as $F=ma$. This is expressed in Newtons (N), the force required to accelerate one kg by one m s^{-2}.

This might sound somewhat perplexing but consider *weight*. This is an expression of force which uses the familiar property of gravity as the source of acceleration. So the equation for weight is easily rewritten as $F_g=mg$.

An astronaut in orbit has a weight of zero yet might have a mass of 85 kg. Likewise, when a boxer takes a jab at his opponent, the fist's mass remains constant, but the force at impact is greater if the fist moves faster, producing greater acceleration at impact.

Now we know what we need to understand a very familiar meteorological property: *pressure*. Pressure is force per unit area: $P=F/A$, where F is force (N) and A is area (m^2). So if we put the units of measurement where F and A appear, it is clear that we will be calculating N/m^2. Conveniently, this happens to be the definition of a Pascal (Pa). If we divide any value of Pascals by 100, we end up with hectopascals (hPa), which are the same as millibars (mb). Those are the key units of pressure in operational meteorology and we refer to them interchangeably in this book.

Pressure may also be measured in terms of the height of a column of mercury in an evacuated tube, and in such a case is expressed in millimeters (mm) or the much more familiar inches of mercury (in Hg). One inch of mercury equals 33.8636 mb (hPa).

1.5. Temperature

Temperature, *T*, is an expression of the amount of kinetic energy in the molecules of a substance. In meteorology, this manifests itself as heat. The standard unit of temperature is the Kelvin (K). When the temperature is zero Kelvin, no molecular energy is present. Another reference point for this scale is the melting point of water, which is 273.16 K. This constant, 273.16 K, is highly recommended for practicing meteorologists to memorize.

Temperature is also measured in degrees Celsius, which establishes 0 °C as the melting point of water and 100 °C as the boiling point in a standard atmosphere. The intervals of Celsius degrees, fortunately, are the same as those of Kelvin, so it follows that if water melts at 273.16 K, it boils at 373.16 K.

Wet bulb

Humidity is generally measured by determining the rate of evaporation of water into the air. This is generally accomplished by the use of a psychrometer, an instrument consisting of two thermometers, the bulb of one being surrounded by a moistened cloth from which water is free to evporate and the other being freely exposed to the air. Due to evaporation, the temperature indicated by the thermometer with the "wet-bulb" will be somewhat lower than the other, the more rapid the evaporation the greater the difference between the two thermometers. When the air is saturated the two thermometers will indicate the same temperature, while if it is very dry the wet-bulb thermometer may indicate a temperature many degrees lower than the other. By the use of suitable tables this difference in temperature may be used in calculating the humidity.

GEORGE TAYLOR
"Aeronautical Meteorology", 1938

The final unit of temperature is degrees Fahrenheit. In the grand scheme of things this is a somewhat outdated system, but it remains in very heavy use in the United States. The reference points of 0 °F and 100 °F were based hundreds of years ago on chilled brine and body temperature, respectively. In a standard atmosphere the melting and boiling points of water in the Fahrenheit system are 32 °F and 212 °F.

1.6. Density

Density (abbreviated by the Greek letter rho, ρ) describes the mass of a *specific volume* of a substance: $\rho=m/v$. It is given in kilograms per cubic meter (kg/m³).

For meteorologists, the ideal gas law relates density to temperature. One expression states that $\rho=P/(rT)$, where r is a constant for dry air, so we can simplify this to the statement "ρ is proportional to P/T" ($\rho \propto P/T$). Examine this closely. Simple algebra shows that if we hold P constant and raise T, then ρ will decrease. This means that cold air is dense and warm air is not.

Density is of particular interest to pilots, since low density means that a plane must move faster to achieve the same airspeed. The McDonnell Douglas DC-10 Series 15, the Boeing 707 Series 220, and the Vickers VC-10 are examples of multimillion dollar airplanes which were designed specifically for "hot high" airports such as Mexico City, Denver, and Johannesburg. Low station pressure and high temperatures all conspire to produce low $\rho=P/T$ values, and the result is that an aircraft requires more power or a longer takeoff roll to get airborne than it would otherwise. The main feature of these aircraft models, not surprisingly, is engines which are more powerful than would normally be expected.

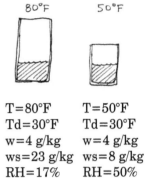

80°F	50°F
T=80°F	T=50°F
Td=30°F	Td=30°F
w=4 g/kg	w=4 g/kg
ws=23 g/kg	ws=8 g/kg
RH=17%	RH=50%

Figure 1-1. Simplistic *analogy* of how moisture variables change when the temperature changes. In this example, a parcel of air is represented by a glass of water. The parcel contains 4 g/kg of water vapor. When we cool the air, the so-called "holding capacity" of the glass shrinks. The saturation mixing ratio decreases and the relative humidity goes up. Although this is excellent for providing a quick grasp of how basic moisture properties are interconnected, forecasters should dismiss the metaphor of the parcel actually having "holding capacities" or the qualities of a drinking glass.

Figure 1-2. The Vickers VC-10 was specifically designed for operations in low-density air, serving the proverbial "hot/high" airports such as Johannesburg and Nairobi. The low density values mean that the wings and engines are less efficient, and planes must move faster to achieve the same performance figures. The United States developed special "hot, high" models of its heavy jet planes, which included the Boeing 707-220 and the McDonnell Douglas DC-10 Series 15. Most of these planes have retired from passenger service, and today's fleets use more powerful engines and have better safety margins in low-density air. *(BAE Systems)*

1.7. Water

All matter is made up of three main phases. In order of increasing energy state, these are solid, liquid, and gas. Matter which has a temperature below its melting point will tend to remain in the solid state and that above its boiling point will change to gaseous state. That said, liquid does not need to reach its boiling point to change to a gas because there are always individual molecules that are spontaneously in a high energy state and as a result will change to a gas. This is process is known as evaporation.

The Earth's air is made up of nitrogen, oxygen, argon, and other gases. These always remain in the gaseous state since their boiling point is around –200 °C. Water, however, spans the whole range of phases. In its gaseous form, it is known as *water vapor* — not to be confused with droplets, which are liquid.

Several different measurements for moisture are commonly used in meteorology:

1.7.1. MIXING RATIO (w, g/kg) is simply the ratio of the mass of water vapor to the mass of dry air in a given volume. The greater the mixing ratio, the more water contained in a given volume of air. On a summer day in Florida, a typical mixing ratio will be about 12 g/kg.

1.7.2. SATURATION MIXING RATIO (w_s, g/kg) represents the *maximum* possible mixing ratio for a given parcel of air. If the mixing ratio reaches this value, the air is considered to be

Hydrostatic equilibrium

Hydrostatic equilibrium not only determines the structure of the earth's atmosphere with respect to gravity. It is also used by astrophysicists to describe the balance of a star's energy with respect to gravity. If the energy output is the same over a given period, the star will neither expand nor contract, and it is said to be stable.

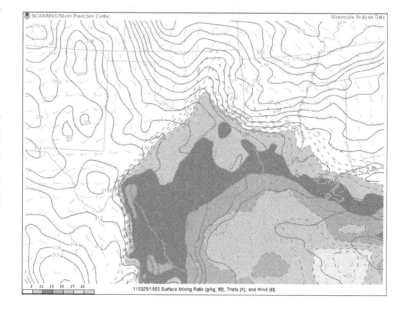

Figure 1-3. Mixing ratio (shaded region and light dashed lines) provides a useful proxy for dewpoint. Furthermore since mixing ratio increases exponentially with a given dewpoint increase, it tends to highlight moisture better in severe weather situations, providing more contours at higher moisture values and less of them in drier air. This example is taken from the SPC Mesoanalysis web site. (25 March 2011 / 1800 UTC)

saturated, and water vapor must condense into liquid droplets or ice crystals if suitable condensation nuclei are present. Saturation mixing ratio is a largely a function of *air temperature.*

1.7.3. RELATIVE HUMIDITY (percent). Astute readers might notice a clear relationship between mixing ratio and saturation mixing ratio. Indeed, if w is 12 g/kg and w_s is 24 g/kg, then we can say the air has half the water vapor it needs for saturation to occur. This ratio is known as *relative humidity* and is equal to w/w_s. It can be seen that if air cools, it will eventually reach a point where w_s falls to w. At this point, the relative humidity is 100% and saturation occurs.

1.7.4. DEWPOINT TEMPERATURE (T_d) is the temperature at which saturation will occur if the air is cooled, omitting any change in pressure. When this point is reached, water vapor (gas) will change to liquid form if suitable condensation nuclei are present, and droplets will form on these nuclei. In the atmosphere, this results in clouds and precipitation. Dewpoint temperature is proportional to the amount of water vapor in the air, so it makes a good indicator of absolute (actual) moisture.

1.7.5. DEWPOINT DEPRESSION is simply the difference between the actual air temperature and the dewpoint temperature, $T_{dd} = T - T_d$. The lower the value, the closer the air is to saturation and the higher the relative humidity. A dewpoint depression of less than 5 C° at a given level in the atmosphere is considered to be suitable for the formation of clouds.

1.7.6. WET BULB TEMPERATURE (T_w) is the lowest temperature to which air can be cooled by evaporation. It can be directly measured using a wet-bulb thermometer, aspirating a wet wick and reading the lowest measured temperature. The wet bulb temperature always lies between the values of temperature and dewpoint.

1.7.7. VIRTUAL TEMPERATURE (T_v) is an expression of temperature that takes into account the density of water vapor. Water vapor is over a third less dense than dry air, so adding it to dry air reduces its density. Virtual temperature is simply the temperature of a hypothetical mass of dry air whose density equals that of the sample containing water vapor. The formula is $T_v = T + (w/6)$, where T is degrees Celsius or Kelvin. It can be seen that by increasing the moisture or dewpoint, T_v will be as much as several degrees higher than the ambient air temperature. This increases the buoyancy of the air.

1.8. Coriolis force

The Coriolis force is an apparent force that acts on objects moving across Earth's sphere. It occurs because of the change in angular velocity with latitude when viewed within an Earth-based rotating coordinate system. The Coriolis force produces a rightward deflection on any moving air parcel in the northern hemisphere (leftward in the southern hemisphere), with a magnitude directly proportional to its velocity.

The Coriolis force equals fv. The term f is $2\Omega \sin(\phi)$, where Ω is the Earth's angular velocity (a constant), and ϕ is the latitude. Therefore f is zero at the Equator and increases with latitude. The term v is the velocity of the parcel's motion.

The above definition shows that the Coriolis force is zero when the parcel is stationary. When the parcel is in motion, Coriolis force is directly proportional to the velocity of the air parcel. The Coriolis force also increases in strength as one moves away from the Equator.

1.9. Wind forces

Pressure gradient is defined as $\Delta P/\Delta n$ — change of pressure observed across a specific change in distance. This is always measured perpendicular to the isobars. An obvious example of a pressure gradient is in science fiction movies when a character "blows the hatch" on the airlock. When the door opens, the pressure gradient in the airlock is about 1000 mb over a distance

Figure 1-4. A parcel in geostrophic balance in the northern hemisphere, not affected by friction or other outside forces. We assume the parcel is already in motion. The parcel will be attracted towards the lower pressure in the +n direction. This is resisted by Coriolis force in the −n direction. The amount of force in the +n direction is proportional to the strength of the pressure gradient (isobars shown as a series of solid lines beneath the parcel). The amount of Coriolis force in the −n direction is proportional to the parcel's velocity.

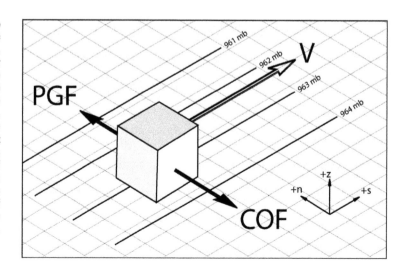

of 5 to 10 meters. In the equation $\Delta P/\Delta n$, the high pressure gradient ΔP divided by a short distance of Δn makes the pressure gradient value quite large.

The pressure gradient force (PGF) defines the force causing the air to respond to pressure gradient in order to "fill the void". The pressure gradient force equals $-(1/\rho) \times \Delta P/\Delta n$., with ρ signifying the air's density. In short, the pressure gradient force is largely proportional to the pressure gradient itself. Not surprisingly, in *Aliens* when Sigourney Weaver's character ejected a monster in an airlock, the extremely high pressure gradient resulted in a very strong wind directed toward the vacuum of space. The contour gradient force (CGF), $\Delta z/\Delta n$, is the measure of pressure gradient on upper-level charts and functionally is the same thing as pressure gradient force.

1.10. Geostrophic wind

Geostrophic wind is an ideal, imaginary wind that would result if an exact balance between the Coriolis force and the pressure gradient force existed. It can easily be calculated by analyzing only the existing pressure or contour gradient and taking into account the latitude. It assumes straight-line flow, with no centrifugal force, and the absence of friction.

If the actual wind is flowing faster than the geostrophic wind, the Coriolis force becomes dominant and forces the air to flow towards higher heights (where it loses kinetic energy) until the velocity slows down and thus the Coriolis force subsides. This

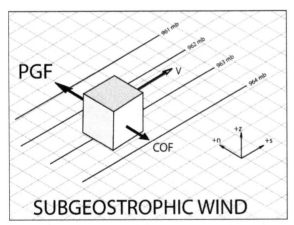

SUBGEOSTROPHIC WIND

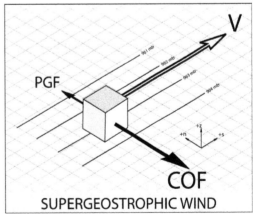

SUPERGEOSTROPHIC WIND

Figure 1-5. Ageostrophic flow in the northern hemisphere. If the parcel's velocity is lower than the geostrophic wind for the given pressure gradient, the contribution from the Coriolis term will be reduced and this will result in the pressure gradient force becoming the dominant force. As a result, the parcel tends to move toward lower pressure, and the velocity vector moves slightly to the left of the V vector shown here. This condition is called subgeostrophic wind. On the other hand, if the parcel's velocity is higher than the geostrophic wind for the given pressure gradient, the contribution from the Coriolis force is amplified and the parcel tends to move toward higher pressures. In the example above, the velocity vector will adjust somewhat to the right. This condition is called supergeostrophic wind.

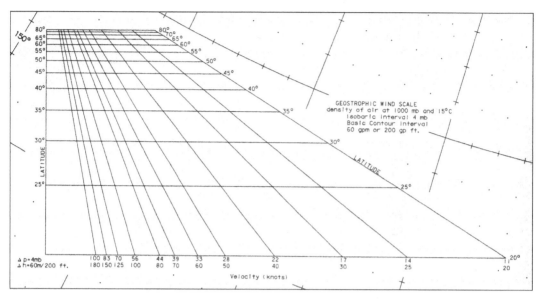

Figure 1-6. Geostrophic wind scale as printed on a US Department of Defense weather plotting chart. These wind scales were common on weather charts during the 20th century, especially those that were disseminated to merchant marine fleets over radiofax. Using any 4 mb pressure difference, commonly that between two isobars, the gap can be compared to this scale and a geostrophic wind speed can be determined. Due to the increasing strength of the Coriolis force with latitude, the scale is graduated according to latitude. A given pressure gradient results in less wind at the poles than in the tropics.

Coriolis force? Coriolis effect? Is it the Coriolis force that deflects a parcel, or is it the Coriolis effect? Technically, it is an effect. However, all motions in meteorology are measured using the Earth's rotating coordinate system. Within this system, the Coriolis effect does in fact behave as a force. However, the Coriolis force can only change the direction of a mass, not its kinetic energy, so it does not do work.

condition is known as a supergeostrophic wind. This will be an important concept as we learn about jet streaks later, since the exit (usually, the east) region of a jet streak tends to be supergeostrophic.

If the actual wind is flowing slower than the geostrophic wind, the pressure gradient force becomes dominant and forces the air to flow toward lower heights (where it gains kinetic energy) until the velocity speeds up and the Coriolis force becomes more dominant. This condition is known as subgeostrophic wind. Again, it is an important concept that will be explained in the section about jet streaks, since the wind on the entrance (usually, the west) side is usually subgeostrophic.

1.11. Gradient wind

Finally there is friction. This is a force that opposes motion. In terms of direction, it always acts opposite to the direction of v (velocity). If friction increases, velocity decreases. Since friction always opposes the velocity, its presence leads to a reduction of the Coriolis force. Therefore when friction is added, the pressure gradient force dominates Coriolis force, helping the parcel turn more directly toward lower pressure. This is why winds flow more directly into low pressure over land than over oceans. It also explains why winds tend to back and slow down as one moves downward toward the Earth's surface, where friction is occurring.

The gradient wind is the wind that would result of there was a balance between the pressure gradient force, the Coriolis force, and centrifugal force. It assumes no friction.

In a supergradient wind, the Coriolis force (acting to the right in the northern hemisphere) balances against a combination of the pressure gradient force and centrifugal force (acting to the left). Since the centrifugal force will help the parcel accelerate into lower pressures (increasing its kinetic energy), it speeds up until the Coriolis force strengthens. Therefore the wind is stronger in anticyclonic flow than what the geostrophic wind would indicate.

In a subgradient wind, the Coriolis force and centrifugal force combine to balance against the pressure gradient force. Since the centrifugal force helps the parcel move away from lower heights (reducing its kinetic energy), it slows down until the Coriolis force weakens. Therefore the wind is weaker in cyclonic flow than what the geostrophic wind would indicate.

1.12. Cyclostrophic wind

Any type of curved flow adds another force: centrifugal force. This is an apparent force that deflects particles away from the center of rotation. It is always directed outward from the axis of rotation and is perpendicular to the direction of v (velocity). In systems with a high Rossby number, the Coriolis force is weak and centrifugal force is a much more important part of the balance of forces. Circulations with high Rossby numbers are always found in the tropics, where the Coriolis force is weak, or in very small-scale circulations with relatively intense velocity.

When this condition occurs, the balance of wind forces are cyclostrophic, with centrifugal force resisting the pressure gradient force. The winds in hurricanes and typhoons are often near cyclostrophic balance, as are tornadoes and dust devils. This balance of forces is why the systems don't fill immediately and dissipate. Centrifugal force preserves the circulation. Either clockwise or counterclockwise flow is possible.

Systems with low Rossby numbers don't just extend just to intense wind systems; small-scale ocean circulations and even the water in a flushed bathroom toilet can be described as having a high Rossby number and exemplifying cyclostrophic flow.

1.13. Vorticity

Vorticity defined is a measure of spin of the air, measured in radians per second. It is created by either shear, curvature, or a combination of the two. If we picture whirls of air forming between the cars on a freeway, it's easy to see that they will form wherever a lane of fast traffic is adjacent to a lane of slow traffic, but not when both lanes are the same speed. This demonstrates

Tidally locked worlds
Science-fiction authors and fans often speculate what the Earth would be like if it stopped rotating. Would the sunward side become an oven with a frozen night side?

Jason C. Goodman, an experienced climatology graduate student, speculated that the loss of the Earth's rotation would vastly improve the efficiency of the atmosphere as an engine to dispel temperature differences. In fact, the temperature difference between the night and day side might be as little as 10 degrees C. However between these zones the atmospheric engine would rage in full force, with a 500 mph windstorm transporting cold air from the night side to the day side, and aloft a strong jet stream transporting warm air to the night side. The livable regions would be the center of the day side, where light winds, torrential rain, and warm weather would predominate, and on the center of the night side, where light winds, clear skies, and cool weather would be found.

Several physicists tended to agree with the assessment, figuring a night temperature near freezing and a dayside temperature just above 130 deg F, but with much lighter winds between the two zones.

Still interested? Check out "Simulations of the Atmospheres of Synchronously Rotating Terrestrial Planets Orbiting M Dwarfs: Conditions for Atmospheric Collapse and the Implications for Habitability" by Joshi, Haberle, and Reynolds (Icarus V129, pp 450-465, 1997).

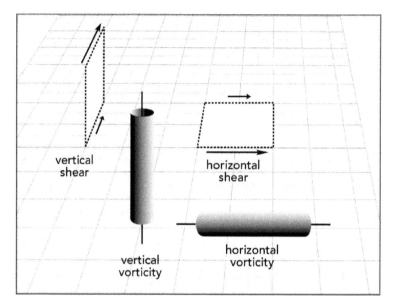

Figure 1-7. The two primary types of vorticity: vertical vorticity, that is, vorticity around a vertically oriented axis, and horizontal vorticity. For the most part, meteorologists are interested in vertical vorticity except when assessing shear between two layers in the atmosphere. As can be seen here, it is vertical shear which contributes to horizontal vorticity.

vorticity arising from shear. Likewise, if the eddy is carried in uniform flow around a bend in the road, the bend will impart spin to the eddy. This is vorticity from curvature.

In meteorology, circulations of interest are usually on the order of 10^{-5} s^{-1}, so this is often abbreviated informally to a *unit of vorticity*, which equals 10^{-5} s^{-1}.

Being a measure of spin, vorticity has an axis and subscribes to the *right hand rule*. When you hold out your right hand and your curled fingers point to the direction of spin, your upstretched thumb represents the axis of spin. If we consider horizontal tubes of vorticity, like a roll cloud or pencil rolling down a slope, the axis points horizontally and thus it is considered horizontal vorticity. Vertical forms of vorticity include whirlpools or merry-go-rounds.

Nearly all expressions of vorticity in operational meteorology refer to vertical vorticity unless specified otherwise. Horizontal vorticity is limited mostly to evaluating shear through a layer, such as for forecasting severe storms, turbulence, and cloud patterns.

Referring to the right-hand rule again, it can be seen that positive vorticity (in the direction in which the fingers point) represents cyclonic [anticyclonic in SH] spin. Locations with the highest amount of positive vorticity are by convention marked with an "X". Likewise, negative vorticity has anticyclonic [cyclonic in SH] spin, and the highest intensities are marked with an "N".

There are also two types of vorticity used in operational meteorology. Relative vorticity, ζ, is vorticity resulting purely from both shear and curvature. This is typically used in mesoscale charts and products showing the low-level wind fields. When the rotation of the earth is added in, we get an expression known as absolute vorticity ($\eta = \zeta + \Omega$). The Earth adds a latitude-dependent amount of vorticity that ranges from 0 units to 14 units at the North Pole. Nearly all upper level charts in routine forecasting use absolute vorticity, while surface wind diagnostics tend to use relative vorticity.

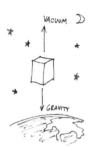

1.14. Horizontal coordinate systems

As map makers, mathematicians, and computer gamers know, the most basic coordinate system is the Cartesian system, which describes locations in three-dimensional space using x, y, and z position. In this scheme, the x axis points to the east, the y axis points to the north, and the z axis points into space. So locations in the $+x$ direction are considered to be toward the east.

Compass directions do not mean much to the balance of forces on an air parcel. What matters is usually whether forces are acting perpendicular or parallel to the parcel motion. To address this issue, the natural coordinate system views coordinates from the parcel's frame of reference. Assume we are a ship's captain standing at the helm on a moving parcel, looking straight ahead like a ship's captain at the helm. The s (streamline) axis points straight ahead, with areas ahead considered the $+s$ direction or downstream, and areas behind considered the $-s$ direction or upstream. The n (normal) axis points to the left, with the left hand side the $+n$ direction, and the right hand side the $-n$ direction.

The right hand rule is useful to help remember which direction is which. The thumb points toward $+x$ or $+s$, always downstream or east. While doing so, the index finger, pointed perpendicularly from the thumb, will point toward $+y$ or $+n$ and the middle finger, pointed perpendicular along a third axis, will point toward $+z$, up.

1.15. Vertical coordinate systems

When you hear "that plane's at 35,000 feet" or "the height of Elvis' UFO was 700 feet", you are hearing a reference to a height coordinate system. In this system, the z axis points upward. Though height and altitude are the most familiar coordinate systems for the atmosphere, meteorologists frequently use the pressure coordinate system, which describes heights in terms of pressures (mb or hPa).

Hydrostatic equation

$\Delta P / \Delta Z = -\rho g$

Since a vacuum exists above the earth, air masses do in fact "sense" the pressure gradient between Earth and space and have a tendency to rush out into space to fill the void. This doesn't happen, though, because the tendency to of an air parcel to move towards space is offset by the pull of gravity. This balance is called hydrostatic equilibrium, and is illustrated by the hydrostatic equation above. The left term shows that the vertical pressure gradient (change in pressure, P, per change of vertical distance, Z) is a function of the right term, density multiplied by gravity. If gravity was stronger, the left term would increase, suggesting the pressure would decrease much more strongly through a given layer than it would otherwise.

Hypsometric equation

$\Delta Z = k T_v$

This equation indicates the thickness of a layer (bounded by two pressure surfaces) in weather forecasting. The conceptual equation is, where ΔZ is thickness in meters, k is a constant unique to each layer, and T_v is the mean virtual temperature of the layer. This shows that thickness (difference in distance) of a layer is directly proportional to its mean virtual temperature, and is proportional to moisture. As mixing ratio increases, virtual temperature increases and thus thickness increases.

Pressure coordinate systems (isobaric coordinates) are also used. To better explain this system, the 500 mb level represents the height to which a barometer would have to ascend to give a reading of 500 mb. This is usually around 18,000 ft (5 km) above sea level.

When visualized over a broad horizontal area, the height at where this occurs forms an imaginary *pressure surface*. This height varies according to the atmospheric density below that particular point. Forecasters regularly map out the height of pressure surfaces using lines of equal height, also known as contours. Areas of high height indicate that the height of the pressure level is unusually high, and this correlates with high pressure or a warm atmosphere below. Areas of low height are correlated with low pressure or a cold atmosphere below.

Since pressure increases as altitude decreases, the Cartesian system's *z* coordinate is replaced by the *p* (pressure) coordinate when we use a pressure coordinate system. It points downward, not upward.

Figure 1-8. The value of a mesoscale observation network is illustrated here by Texas Tech University's West Texas Mesonet (mesonet.ttu.edu). Around the periphery of the chart is the standard Federal observation network, consisting mostly of ASOS sites at a spacing of about 50 miles between stations. A mesonet like this one provides a spacing of about 10 miles, allowing it to truly sample mesoscale-sized weather features. *(Texas Tech University)*

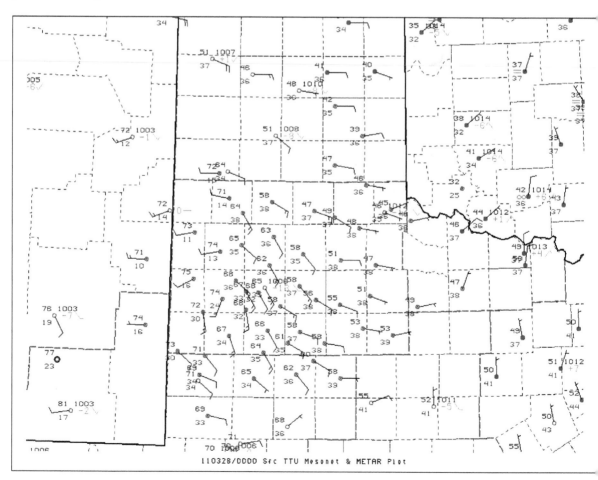

110328/0000 Sfc TTU Mesonet & METAR Plot

1.16. Scale

Weather systems and circulations are widely described in terms of their scale, so all practicing forecasters must be familiar with this terminology. It should be emphasized that the distinctions between each category are not clear-cut and that other scale systems have been put forth. Shown here are the most common ones used in American meteorology.

1.16.1. PLANETARY SCALE. Scales on the order of 10,000 km or greater, or spanning a life cycle of weeks or months, are described as planetary or global. Some examples of planetary systems are the "general circulation" of heat at the equator and cold air at the poles, and the broad system of westerly flow across the globe.

1.16.2. SYNOPTIC SCALE. Systems measuring on the order of 1,000 to 10,000 km, spanning a life cycle of days to weeks, are considered to be synoptic scale. It encompasses baroclinic waves, large frontal systems, and long waves.

1.16.3. SUBSYNOPTIC SCALE. Systems on the order of 100 to 1,000 km in size or having a life cycle of hours to days are considered subsynoptic. It encompasses squall lines, tropical cyclones, and short waves. In practice, the upper end of this scale is often annexed to the synoptic scale regime, with the lower end given to the mesoscale regime (see below).

1.16.4. MESOSCALE. Scales on the order of 10 to 100 km (sometimes defined as high as 1,000 km) encompassing life cycles of minutes to hours fall into the mesoscale realm. It includes thunderstorms, cloud clusters, thunderstorms, coastal fronts, drylines, orographic effects, and so forth.

1.16.5. MICROSCALE. Scales on the order of 10 km or less or having a life cycle of seconds to minutes are considered microscale. This includes turbulence, small clouds, tornadoes, dust devils, mountain waves, etc.

1.17. Atmospheric structure

There are four primary atmospheric layers that concern meteorologists: the troposphere and stratosphere, and the boundary layer and free atmosphere. Though the mesosphere and thermosphere are important parts of the Earth's atmosphere, they do not have a day-to-day influence on forecast problems.

Vertical speeds
Typical time it takes to go 1 meter (3.2 feet) up or down

UPWARD

Severe storm updraft	0.01 sec
Cumulus updraft	1 sec
Air mass on rainy day	20 sec

DOWNWARD

Air mass on fair day	3 min
Cloud droplet	1.5 min
Large cloud droplet	4 sec
Rain drop	0.25 sec
Golfball-sized hail	0.04 sec

Weight of a cloud
A cloud's weight depends on the distribution of water and ice. Typical densities range from about 0.3 to 0.7 g/m^3 in a benign weather cloud to 10 g/m^3 in a typical thunderstorm. Assuming a cumulus cloud has a density of 0.5 g/m^3 and has a size of 1 km^3 (1 billion cubic meters), we can calculate that it has a weight of 500,000 kg, or about 550 tons. This is a little heaver than the weight of a fully loaded Boeing 747.

1.17.1. TROPOSPHERE. The troposphere is the lowest atmospheric layer and contains nearly all of the Earth's weather. The temperature normally decreases with height within the troposphere. The top of the troposphere is called the *tropopause*. Its height is largely a function of the mean temperature in the troposphere, and ranges from about 4 miles (7 km) in polar regions to 11 miles (17 km) in tropical regions.

1.17.2. STRATOSPHERE. The stratosphere is a layer above the troposphere where the air temperature remains constant or increases with height. It extends from the tropopause up to the stratopause at a height of about 30 miles (50 km). The temperature increase with height is caused by a layer of ozone in the stratosphere which absorbs energy from incoming solar radiation. Although the stratosphere is not important to the daily forecast problem, pockets of warm air in the lower stratosphere (called "warm sinks") are often found and tend to be associated with regions of strong upward motion in the troposphere. Nacreous clouds, also known as "mother of pearl" clouds, occasionally occur in polar regions at a height of 10 to 15 miles (15 to 25 km), however they are currently of no concern to operational meteorologists.

Figure 1-9. Atmospheric layers stand out like pancakes in this shot of the horizon from Space Shuttle Columbia over Baja California in October 1995. Research satellites have begun using a technique called "limb viewing" which looks at these layers on the horizon to measure ozone concentration. *(NASA)*

1.17.3. PLANETARY BOUNDARY LAYER (PBL). The boundary layer is that part of the atmosphere where friction from the Earth's

surface affects air motion. The top of the boundary layer (called the gradient level) is usually 2300 ft (700 m) but is lower over the oceans and higher over mountainous terrain. Higher wind speed and higher instability tend to increase the gradient level due to increased turbulence and mixing within the boundary layer. The balance of forces in the PBL are best described by the gradient wind.

1.17.4. FREE ATMOSPHERE. The free atmosphere is that layer of the earth's atmosphere above the boundary layer. It is normally found from about 3000 ft MSL upward. Within the free atmosphere, the wind is largely within geostrophic balance.

1.18. Global circulation

The global circulation is a model that attempts to describe the planetary motion of air. In an simplified atmosphere, these patterns would dominate the weather, but it is now known that two of the cells that make up the three-cell global circulation model presented here are usually dominated by complex physical processes and are rarely a factor in everyday forecasting.

1.18.1. HADLEY CELL. This concept, developed in 1735 by meteorologist George Hadley, describes the heat-driven circulation that originates at the equator. Solar heating heats

Figure 1-10. Thunderstorms in the tropics help drive the Hadley Cell, with hundreds of deep convective cells active at any given moment. This image was photographed by Space Shuttle STS-35 over Vietnam on 25 July 1999, showing vast anvil plumes diffusing tremendous amounts of moisture throughout the troposphere. *(NASA)*

the atmosphere directly on the equator. This air rises and flows toward the poles. It was originally thought that this air went all the way to the poles, forming the overall global circulation, but it was later demonstrated that the Coriolis force begins deflecting the air (rightward in NH, the northern hemisphere), where it accumulates in the subtropical latitudes. This forces the air to sink toward the surface. The result is a band of sinking air at approximately 30 degrees latitude, generally called the subtropical high. Air sinks to the surface then flows back toward the equator. It is deflected rightward (NH), resulting in a flow of prevailing winds moving from northeast to southwest (NH). This wind is known as the trade winds or the tropical easterlies. It explains why winds usually blow from the east in tropical locations.

1.18.2. FERREL CELL. This is a concept that describes the circulation that exists between 30 deg latitude and approximately 60 deg latitude. It was proposed in 1856 by American mathematician William Ferrel (1817-1891). Surface air originating from the sinking region of the Hadley cell flows poleward, deflecting (rightward in the NH) as it moves. The result is a band of prevailing winds that flow from southwest to northeast (NH) at the surface in mid-latitudes. This wind is known as the prevailing westerlies. If the Ferrel Cell concept

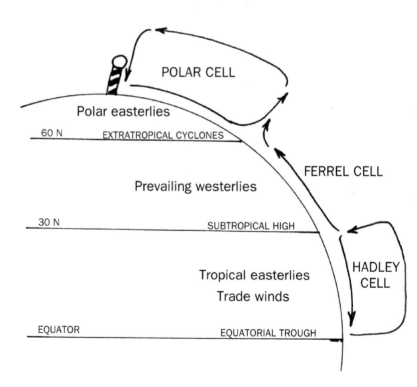

Figure 1-11. Simple model of the Earth's global circulation. The earth is divided into three main circulation cells and three main bands of wind.

is taken literally, the air would rise at about 60 deg latitude and return equatorward, deflecting right (NH) to form an easterly upper-level wind, but this does not actually occur. The Ferrel cell concept is an oversimplification, but it is relevant in describing the "prevailing westerlies" at the surface.

1.18.3. Polar cell. This cell describes the conceptual circulation that exists between the north pole, where excess radiation of heat into space continually creates a cold surface air mass, and a belt at roughly 60 degrees latitude. Cold air from the arctic regions flows south and is deflected by the Coriolis force at about 60 degrees latitude, where it accumulates. The net surface convergence in this belt results in rising motion. The air rises and diverges aloft, flowing back northward toward the poles. In reality this cell is also an oversimplification since flow in the high latitudes is dominated by lows, highs, and atmospheric waves. However the polar cell does have importance in describing the concept of "polar easterlies" and cold air outbreaks.

Chapter One
REVIEW QUESTIONS

1. If the earth's gravity suddenly decreases by half, describe the change in air pressure. The relevant equations are $F=ma$ and $P=F/A$.

2. The saturation mixing ratio has increased. What does this indicate? Will the relative humidity go up or down?

3. If the mixing ratio of a parcel increases, what is the effect on its density? Which property explains this?

4. The mixing ratio increases. The temperature has stayed the same. How has the dewpoint and relative humidity changed?

5. If friction increases and the wind slows down, how does the balance between pressure gradient force and Coriolis force change?

6. Is the pressure coordinate system proportional or inversely proportional to the height coordinate system?

7. The air at 1 km is from the south at 15 kt while the air at 2 km is from the south at 25 kt. Is vertical or horizontal vorticity is present and which way does its axis point, relative to the wind or to the ground.

8. Which of the hemispheric cells is driven by the sun?

9. Which map scale is most likely used on surface charts worldwide?

10. Do wind barbs point upstream or downstream?

SPEED (knots) (10)	CHARAC-TER (11)	ALST6 (inches) (12)	REMARKS AND SUPPLEMENTARY CODED DATA — DESIRED ORDER OF ENTRY: RVR, SFC based obsc phenomena, remarks elaborating on preceding coded data 3- and 6- hourly additive data, radiosonde data runway conditions, weather modification (15)	STATION PRESSURE (inches) (17)	TOTAL SKY COVER (21)	OBS INIT (18)
06		988	FG WR// FIRST/ 20025 90401*	24.355	10	CS
04		988	FG F BNK NW PSR09 F2 Coz 15.0	24.355	10	CS
			PSR13P WET			CS
04		990	F4 F BNK W-N PSR13P WET/ /// 16//	24.370	10	CS
03		990	F2			TAV
03		990	F2 VSBY LWR SW-NW PSR13P WET	24.375	10	TAV
07		991	F2 VSBY LWR W-N PCPN VRY LGT			TAV
08		992	F1 VSBY LWR W-N PCPN VRY LGT			TAV
08		992	F1 VSBY LWR W-N PCPN VRY LGT PSR13P WET	24.385	10	TAV
			SLR//			TAV
08		993	F1 VSBY LWR W-N			TAV
09		993	S3			TAV
09		993	11000 SLR//	24.400	10	TAV
08		993	S7			TAV
			WR//			TAV
09		993	S7 VSBY NE2NW11/2 WR//	24.400	10	TAV
08		993	S8			TAV
08		992	S4			TAV
08		991	S6 VSBY NE2NW11/2 TWR VSBY 11/2 WET SNW WR//	24.385	10	TAV
10		991	R14VR50 S6 TWR VSBY 11/2 WET SNW)			TAV
10		991	R14VR38 S7			TAV
10		991	R14VR50 S8			TAV
08		991	R14VR60 S8			TAV
07		990	R14VR60 S8 WET SNW/ 708 172/ WR//	24.375	10	TAV
05		990	S5			TAV
05		990	S4 WET SNW WR//	24.370	10	TAV
07		990	S2 WET SNW			TAV
08		991	S3 WET SNW WR//	24.380	10	TAV
07		991	S5 WET SNW			TAV

KTNX 041300Z 30006KT 21/2SM FG OVC010 M01/M01 A2988 RMK F6 WR// FIRST/ 20050 SLPNO 4/001 T10101010 59004=

KTNX 041400Z 29004KT 21/2SM FG OVC012 M01/M02 A2988 RMK F6 F BKN NW PSR09 SLPNO T10051016=

KTNX 041500Z 29004KT 21/2SM FG OVC011 00/M02 A2990 RMK F4 F BKN W-N PSR13P WET/ SLPNO 8/6// T00001016 5/0//=

KTNX 041526Z 27003KT 5SM FG SCT003 OVC015 A2990 RMK F2 SLPNO=

KTNX 041600Z 30003KT 6SM FG SCT003 BKN015 OVC025 00/M02 A2990 RMK F2 VSBY LWR SW-NW PSR13P WET SLPNO T00001016=

KTNX 041614Z 29007KT 5SM -DZSN FG SCT003 BKN017 OVC025 A2991 RMK F2 VSBY LWR W-N PCPN VRY LGT SLPNO=

KTNX 041649Z 29008KT 5SM SG FG BKN020 OVC027 A2992 RMK F1 VSBY LWR W-N PCPN VRY LGT SLPNO=

KTNX 041700Z 29008KT 5SM SG FG BKN020 OVC027 00/M01 A2992 RMK F2 VSBY LWR W-N PCPN VRY LGT PSR13P WET SLPNO T00001010=

2 OBSERVATION

In today's Internet age, observation is commonly seen as something best left to contractor equipment and complicated electronics. However the seeds of meteorology are planted deeply in the craft of weather observation. There is no way a forecaster can develop skill and expertise without understanding of how the fundamental quantities of meteorology are determined and where in the process errors might creep up. This chapter covers this elemental point of the forecasting process.

2.1. Observation networks

Weather observations are taken daily every 1 to 6 hours at about 5,000 stations around the world and shared via data networks under the oversight of the World Meteorological Organization and International Civil Aviation Organization. Because of the close kinship of meteorology with aviation, it comes as no surprise that a substantial number of weather stations are located at airports and are managed by aviation authorities. This is especially true in the world's wealthier countries, including the United States, Canada, and Europe. In other countries, local weather facilities, operated by the country's weather agency or military organizations, are responsible for weather observations.

2.1.1. DATA DISTRIBUTION. Weather data around the world is fed via radio, phone, or computer into the Global Telecommunication System, a weather network that has a proud history. Its roots go back to 1951, when the World Meteorological Organization was founded as a United Nations agency. The rewards came within a few years as member nations rapidly agreed on a common set of reporting standards and established an international telecommunications network. It allowed the free flow of weather data via teletype. Feeding the network was a system of aviation teletype circuits to obtain reports from airports worldwide. The U.S. Air Force and U.S. Navy also played an integral part in the process, collecting significant amounts of data from third-world countries through its global radio intercept network and sharing this with the WMO network.

2.1.2. ICAO LOCATION INDICATORS. Many hobbyists and professionals are familiar with the 4-letter station codes such as KSFO for San Francisco and EGLL for London. These codes were developed in the mid-1950s by the International Civil Aviation Organization, which standardizes and publishes the codes but leaves the actual assignments up to each member nation. The

Title image
The official weather record for Tonopah Test Range, Nevada on January 4, 1991. The top form shows a Form 10, used throughout the federal government for airways weather observations. The bottom image shows the "longline" teletype version, which was initially transmitted to the world in SAO format, stored at NCDC as rows of tabular records, then converted years later to METAR format. All entries except the top four rows were made by the author. *(Tim Vasquez)*

METAR FORMAT COVERAGE

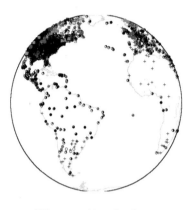

Western Hemisphere

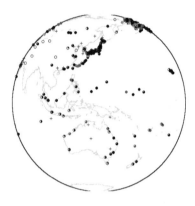

Eastern Hemisphere

SYNOPTIC FORMAT COVERAGE

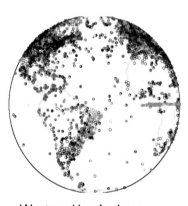

Western Hemisphere

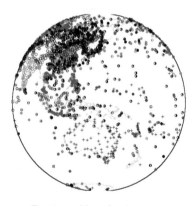

Eastern Hemisphere

Figure 2-1. Surface data availability for October 30, 2000 at 0000 GMT. It's easy to see that METAR coverage is sporadic worldwide except in North America, Europe, and Japan. However coverage with synoptic data is fairly uniform worldwide, though not as dense as METAR in North America. Maps generated by Digital Atmosphere (www.weathergraphics.com).

first one or two letters always represents the country or region; for example, a conterminous U.S. identifier is K---, a Canadian identifier is C---, and a British identifier is EG--. Almost all ICAO identifiers are assigned to aviation facilities.

2.1.3. FAA LOCATION IDENTIFIERS. In the United States, a system of 3-letter airport identifiers was developed in the 1940s by the Civil Aeronautics Administration. These came into widespread operational use starting in the 1960s and are now known as Federal Aviation Administration location identifiers. In the conterminous United States, where ICAO identifiers begin with K, they are generally equivalent to the second, third, and fourth letters. A parallel system exists in Canada, known as Environment Canada identifiers, and are usually compatible with the FAA system. Canada uses identifiers starting with W, X, and Z and the U.S. traditionally avoids use of these blocks.

2.1.4. WMO STATION INDEX. The World Meteorological Organization developed a system of 5-digit numerical identifiers in 1948 which are widely used within meteorology. This was done to resolve growing chaos caused by dozens of station code systems and weather dissemination formats that varied from country to country. Much like the ICAO system, the first two digits identify the country or region and are known as *block numbers*. The conterminous United States has a block number of 72 and 74, Canada uses 71, and Great Britain uses 03. Meteorologists frequently come into contact with these codes when dealing with soundings, radiosonde observations, and synoptic observations. Also many weather stations around the world which are not co-

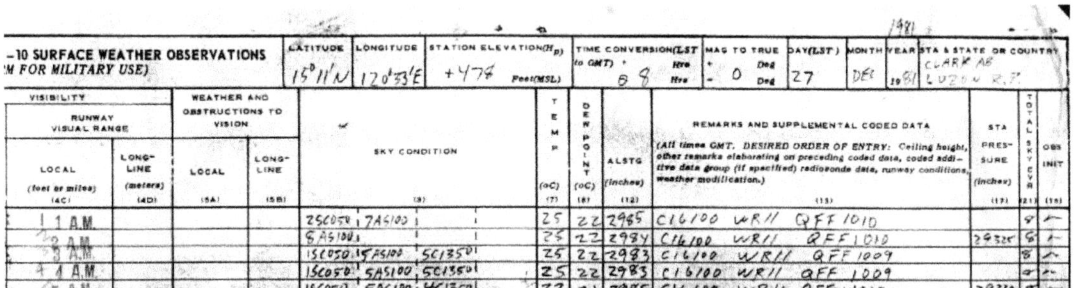

Figure 2-2. Weather observations in the United States are documented on Form 10's, like these shown here, then transmitted via government data networks. Automated observations are usually disseminated directly into the network without human intervention.

Figure 2-3. An Air Force meteorologist at Bishkek, Kyrgyzstan monitors storms moving across the airfield in 2006. Though most of the United States has shifted to automated observing systems, humans remain the backbone of observation process in the U.S. military and throughout most of the world. Most observation networks are closely linked to airports and air bases. *(US Air Force)*

located with an airport will have a WMO identifier but not an ICAO identifier.

2.1.5. WBAN IDENTIFIER. One other identifier sometimes used by U.S. weather agencies is the WBAN (Weather Bureau Air Force Navy) system. This is similar in appearance to the 5-digit WMO system but is not equivalent and does not use block numbers or prefixes. As of 2010 the WBAN system is rarely used in operational meteorology except for certain climatological and historic datasets. The primary U.S. agency that routinely uses WBAN identifiers is the National Climatic Data Center (NCDC).

2.2. Observation coding formats

Three main coding formats for surface weather observations are usually encountered in meteorology. They are described briefly here. Additional details on how to decode the data is provided in the appendix.

2.2.1. METAR OBSERVATIONS. METAR is the most common format for hourly surface observations in North America and Europe. At some stations, a METAR observation is transmitted every 20 minutes, and if a significant weather change occurs it is referred to as a SPECI (special) observation. The format is rather readable and is designed mostly for the aviation sector.

Example of a METAR observation for Duluth, Minnesota:
KDLH 161448Z 06005KT 15SM SCT200 28/22 A3019;

2.2.2. SYNOPTIC OBSERVATIONS (SYNOP). This format is used worldwide, and in addition, it serves as the primary form of surface observations outside of North America and Europe. The observations are usually taken every 3, 6, or 12 hours, and the format, designed mainly for meteorologists, consists of blocks of numerical data. Its primary drawback is that it does not allow plain-language remarks and is very difficult to read without experience.

> Example of SYNOP report for Akrotiri, Cyprus:
> AAXX 26031
> 17601 32575 50000 10112 20084 40197 52004 81250

2.2.3. SURFACE AVIATION OBSERVATION (SAO). The SAO, or "SA" format, was widely used throughout North America until the adoption of METAR in July 1996. It is no longer used except by a handful of older automated stations in Canada. The format, which was optimized for slow teletype transmission, combined readability and brevity and used strict adherence to abbreviations.

> Example of an SAO observation for Fort Worth, Texas:
> FTW SA 1650 40 BKN 7RW- 146/73/62/1518G25/990=

2.3. Temperature

Temperature is measured in a sheltered location, usually on a grassy area away from sources of unusual heating and cooling. Temperature is typically measured at a height of 1.5 m (about 5 feet). On cold, calm mornings, cold air tends to accumulate in a layer close to the ground, which largely explains why a temperature of 34°F might be reported on a morning when frost is all over the ground.

2.4. Dewpoint

This value indicates the amount of moisture at the station. It is always observed at the same location as the temperature. For many decades it has been determined using a wet-bulb thermometer, which uses a wetted wick to measure evaporational cooling, and the reading is then converted to a dewpoint value. This has been gradually replaced by computerized sensors. One type of sensor in common use, including by United States ASOS sites, uses controlled refrigeration of a small mirror to produce condensation detectable by an infrared beam.

2.5. Wind

Wind is generally measured at a height of 10 meters (30 ft) above ground level. The direction that the wind is coming from, as well as its speed, are measured. The value is generally an average value during a period of the past 2 minutes, however any

During the early 1880's the Canadian Pacific Railway and its telegraph lines were pushed across western Canada allowing the establishment of telegraphic weather reporting stations and many climatological and precipitation reporting stations. In eastern Canada during the summer of 1881 the [meteorological] service began to issue forecasts at midnight so that these might be published in the morning newspapers and also be displayed at telegraph stations as soon as they were opened each morning. Another innovation of the early 1880's was the dissemination of weather predictions by means of display discs attached to railway cars. The signal word to be displayed was telegraphed each day at about 1 a.m. to the railway agents who would change the signs on cars each morning in an attempt to provide a reliable weather prediction service for the farming community along the lines of the railway. However, through neglect, the local train hands did not always keep the signal discs up to date, and this arrangement had to be dropped after a decade or so.

MORLEY K. THOMAS,
"A Brief History of Meteorological Services in Canada," 1971

gusts which have occurred during the past 10 minutes comprise the *gust speed*. Older gusts are known as *peak wind* values.

Proper exposure is especially critical with wind data. A small discrepancy in wind values can significantly alter kinematic analysis fields and result in unbalanced model initialization. Since many anenometers at home weather stations are poorly exposed, often hung below roof levels or within tree canopies, the use of wind data from crowd sourced weather networks like MADIS demands the highest degree of caution.

2.6. Pressure

There is one aspect of meteorology that is poorly understood by many amateur forecasts and even some professionals — pressure measurement. In summary, there are three primary systems in use. A forecaster must understand them thoroughly in order to properly make sense of the data and how it impacts the analysis.

2.6.1. STATION PRESSURE (QFE). Station pressure is the *actual*, uncorrected pressure reading. A barometer that reads correctly at sea level and is brought to some other elevation, without adjustment, is displaying the station pressure. While it is not directly used in forecasting, it is the building block of all other

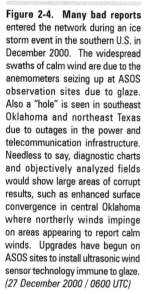

Figure 2-4. Many bad reports entered the network during an ice storm event in the southern U.S. in December 2000. The widespread swaths of calm wind are due to the anemometers seizing up at ASOS observation sites due to glaze. Also a "hole" is seen in southeast Oklahoma and northeast Texas due to outages in the power and telecommunication infrastructure. Needless to say, diagnostic charts and objectively analyzed fields would show large areas of corrupt results, such as enhanced surface convergence in central Oklahoma where northerly winds impinge on areas appearing to report calm winds. Upgrades have begun on ASOS sites to install ultrasonic wind sensor technology immune to glaze. *(27 December 2000 / 0600 UTC)*

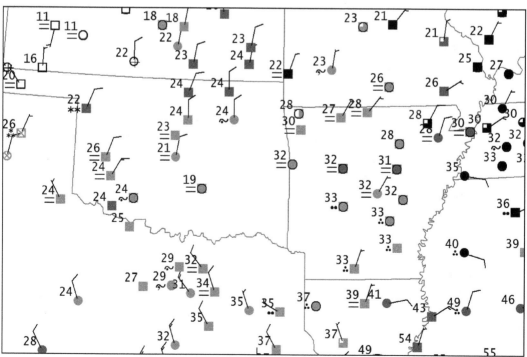

pressure values. Station pressure may be expressed in either inches of mercury or millibars. At stations with a mercury barometer, it may also be expressed in millimeters.

2.6.2. SEA-LEVEL PRESSURE (SLP) is the station pressure "reduced" numerically to sea level. This is done using a correction from a precomputed r-value or r-factor, which in turn is based partly on an average between the current temperature and the temperature 12 hours ago. The intent of this is to remove temperature bias. Sea level pressure is used on a widespread basis on surface maps worldwide. It is expressed in millibars or hectopascals. "Normal" sea-level pressure is 1013.2 mb.

2.6.3. ALTIMETER SETTING (ALSTG, OR QNH) is station pressure value directly reduced to sea level. It is usually expressed in inches of mercury ("in Hg"), though in Europe it may be expressed in millibars. The altimeter setting is the value commonly heard on television and radio weathercasts in the United States and, as its name suggests, it is used for setting aircraft altimeters. When weather hobbyists are setting their home barometer to read pressure, they are normally attempting to obtain altimeter setting, since this factors in elevation but not temperature corrections.

2.7. Visibility

Visibility is a measure of opacity of the atmosphere in a particular direction. It indicates the distance that a normal object under daytime illumination can be seen and recognized. If the visibility is 3 miles, for example, then any objects that are further than 3 miles away are unlikely to be reliably identified.

On most weather observations, a property known as *prevailing visibility* is reported. This is the greatest visibility met or exceeded throughout at least half of the visible horizon. It is given in statute miles or meters. Values of visibility in a specific direction are known as *sector visibility* values and are sometimes included in the remarks of an observation.

2.8. Weather

Weather phenomena may consist of either precipitation, such as rain or snow, or obstructions to vision, such as fog, haze, or smoke. These are always documented in a weather observation. Continuous or intermittent precipitation is distinguished from showery precipitation by the fact that showers are produced by

The first high altitude journey

In the year 1850, M. M. Barral and Bixio conceived the project of ascending to a height of 30,000 or 40,000 feet, for the purpose of investigating certain atmospheric phenomena still imperfectly understood . . . Nothing daunted by the ill-success of their first expedition, and eager to obtain a better result from a second trial, Barral and Bixio determined to ascend again without delay. On July 27, 1850, the filling of the balloon was commenced early in the morning. It proved to be a long operation, occupying till nearly two o'clock. The sky became overcast, and it was past four when they left the earth. They soon entered a cloud at 7,000 or 8,000 feet, which proved to be fully 15,000 feet in thickness. They never, however, reached its highest point, for when at 4 h. 50 m. the height of 23,000 ft was attained, they began to descend, owing to a rent which was then found in the balloon. After vainly attempting to check this involuntary descent, they reached the earth at 5h 3s. The most unexpected result observed in this ascent was the extraordinary decrease of temperature . . . When near the highest point which they attained, their thermometer sank to -38.2°F. The clothes of the observers were covered with fine needles of ice. 'This discovery,' says Arago, in his Report to the Academy, 'explains how these minute crystals may become the nuclei of large hailstones, for they may condense around them the aqueous vapor contained in that portion of the atmosphere where they exist.' . . . The great extent of so cold a cloud explains very satisfactorily the sudden changes of temperature which occur in our climates.

DR. G. HARTWIG
"The Aerial World," 1886

convective clouds, i.e. cumulus or cumulonimbus, and tend to start or stop abruptly.

How do modern automated stations identify weather phenomena? They do so by analyzing scintillation from an infrared beam exposed to the free air. Based on this information, the unit can determine the most likely precipitation or obscuration type. However even the best systems remain highly prone to errors, especially during unusual and "wintry mix" type events.

2.9. Clouds

Like many subjective elements of the weather observation, clouds are not being observed with the same quality that they were before the advent of automated observations. This has been offset somewhat by the increased accuracy and range of cloud height detectors as well as the increase in the number of stations. The proliferation of numerical guidance, high-resolution satellite imagery, and dense observation networks have also marginalized the need for detailed cloud data somewhat. The main elements of cloud data that are traditionally observed are amount, type, and height.

Your hand, when held away from the body and spread out from thumbtip to little fingertip, spans 22 degrees. This corresponds to the refraction angle of light through ice crystals. If cirrostratus is present, put your thumbtip on the sun or moon and look for the halo at the fingertip.

Figure 2-5. Strong vertical shear through a cloud layer can create bizarre and unusual cloud forms, as shown here. *(Tim Vasquez)*

2.9.1. CLOUD AMOUNT. Cloud amount is oktas (eighths). By convention, layer amount is measured based on the principle of summation. According to this principle, all layers of the sky that cannot be seen, due to obstruction by a lower cloud layer, are automatically assumed to be filled with cloud material. So if an observer looks at the sky and can see two oktas of cumulus and three oktas of cirrus is assumed to be hiding behind the cumulus. This means the observation must show five oktas of cirrus.

Automated stations are capable, to a limited extent, of measuring cloud amount. This is usually done by monitoring the vertically-pointed laser beam ceilometer over a long time span and detecting the passage of transitory layers. This is quite similar in principle to staring at the sky through a vertically-pointed paper towel tube and drawing conclusions from what is seen. Recent papers studying the quality of the US observation network (e.g. Dai et al 2006) have found that the ASOS system tends to significantly underestimate cloud amount.

Cloud amounts are commonly expressed either in oktas or in aviation units, with clear (CLR or SKC) indicating a complete absence of clouds and overcast (OVC) indicating total coverage. In between, few (FEW) is used for 1 to 2 oktas, scattered (SCT) for 1 to 4 oktas, and broken (BKN) for 5 to 7 oktas.

2.9.1. CLOUD HEIGHT. In meteorology, a general cloud height refers to *layer height*. This is the base of a cloud layer *above ground level (AGL)*, usually measured to the nearest hundred or thousand of feet. A ceiling may also exist. This is the lowest layer which covers more than half the sky, including any coverage by layers below. Cloud top heights are generally not coded in observations except within PIREP (pilot reports) and radar reports. Cloud height data during the era of human observation, up until the 1990s, relied on a combination of subjective estimates and the use of ceilometers, which use a vertical beam of light or laser energy to obtain information on cloud height. The new ASOS system in use in the United States uses a vertically-pointed laser beam, and tends to be quite accurate assuming all cloud material passes over the equipment.

2.9.3. CLOUD TYPE. Cloud type has made the fastest exodus as an observed property. It used to be a standard part of the METAR report until the 1990s, when it was removed. It is also no longer reported regularly on many observations in North America and Europe. This is partly due to the advent of automated stations, which are completely unable to identify cloud types. Fortunately, the advent of high-quality visible satellite data has made it possible

Cloud nomenclature

For many centuries, it was thought that clouds were too ambiguous to be scientifically identified and classified. However in 1802, Luke Howard, a dedicated English amateur meteorologist, followed the lead of Swedish taxonomist Linnaeus and developed a scheme for categorizing different types of clouds. The scheme used four Latin names "cumulus" (heap), "stratus" (layer), "nimbus" (rain), and "cirrus" (curl), and allowed for mixtures of the words as needed. Howard presented this scheme in his "Essay on the Modification [Classification] of Clouds", presented to London's philosophically-oriented Askesian Society. In the decades ahead it was adopted by meteorologists worldwide.

for meteorologists to remain connected with cloud forms across a forecast area.

Even without the presence of cloud type data in daily observations, cloud types remain one of the fundamental building blocks of a weather forecast and are frequently referred to in journals. They are directly observable on satellite products, they are directly observable by the general public, and the meteorologist sees them outside the office and often spots key processes taking place. Consequently, having a working knowledge of cloud types is absolutely essential even if the data is rarely reported in observations. A full description of cloud types is found in the appendix.

2.10. Upper air systems

Upper air conditions were initially measured with kites in the early 20th century, but gradually radiosonde technology became commonplace by the WWII era. There are also other upper air observing systems in place, which are described here.

2.10.1. RADIOSONDES. Radiosondes are the most important source of "hard" weather data for the upper atmosphere. Twice a day, hundreds of radiosonde stations around the world launch these instruments suspended on hydrogen or helium balloons, and monitor the telemetry as they radio back information on temperature, humidity, pressure, and wind. These are encoded in a code format known as WMO TEMP and disseminated worldwide, where they can be used in soundings and integrated into model forecasts.

2.10.2. AMDAR. Radiosonde stations are expensive to maintain and many countries do not even operate them. Furthermore they are nonexistent over ocean areas. As a result, there has been extensive work to enlist the help of airlines in order to sample data and integrate it into analyses of weather across the globe. While there are very few observations in the lower and middle troposphere, since aircraft do not spend much time at those altitudes, and moisture data is unavailable with these systems, there is extensive upper-air coverage over oceans. This helps provide a fairly accurate picture on the 200 mb and 300 mb charts, where commercial aircraft spend most of their time.

2.10.3. WIND PROFILERS. Wind profilers are low-power radar devices that transmit a signal vertically into the atmosphere and analyze the backscattered signal to determine wind speed and direction throughout the column. The principle behind the system is that the turbulent mixing of air through the column, which yields varying indices of refraction, reveals important properties of the wind. The system operates at 6 kW at 404 MHz, using a flat mesh antenna about the size of a tennis court. The main advantage of this antenna system is that it has no moving parts and requires less maintenance.

Wind profilers have proven to be essential elements in the upper-air network. Their value was proven on May 3, 1999 when a small-scale jet streak not resolved by the rawinsonde network or numerical models emerged over New Mexico and became associated with the development of Oklahoma's worst tornado outbreak.

Often collocated with wind profiler stations are RASS (radio acoustic sounding system) facilities. The signal is generated by a loudspeaker that projects an acoustic sound of about 850-900 Hz upward into the column of air above the station. Needless to say, these devices can be heard around the station, so foam insulation is used to reduce noise pollution. The speed of sound varies with temperature, and this can be sampled to construct a profile of temperature with height. The main drawback is that the vertical resolution of RASS data is nowhere near as good as balloon-launched soundings, and data may be unavailable above 10-15 thousand feet without higher clouds, but the trends over time in each layer can provide valuable information.

The value of RASS has not yet been determined, though it does show some promise in sampling the strength of capping inversions that inhibit thunderstorm development.

NATO Color Codes
Shown here is the NATO color code system, which has been widely used throughout Europe. The visibility and lowest scattered layer are each evaluated, and the lowest color code (the one closest to RED) that was obtained constitutes the NATO Color Code.

Lowest scattered layer
2500 ft or more	BLU
1500-2499 ft	WHT
700-1499 ft	GRN
300-699 ft	YLO
200-299 ft	AMB
199 ft or less	RED

Visibility (meters)
8000 m or more	BLU
5000-7999 m	WHT
3700-4999 m	GRN
1600-3699 m	YLO
800-1599 m	AMB
799 m or less	RED

Visibility (equiv. in statute miles)
5 sm or more	BLU
3 1/2 to 4 sm	WHT
2 1/2 to 3 sm	GRN
1 to 2 1/4 sm	YLO
1/2 to 7/8 sm	AMB
7/16 sm or less	RED

Chapter Two
REVIEW QUESTIONS

1. If you are forecasting weather for Africa, would it be better to use METAR reports or SYNOP reports?

2. The TV show *Survivor* is calling from Guadalcanal Island and needs weather data. Using an online resource, determine the best ICAO and WMO identifier to use to assist them, and indicate which resources helped you find this information.

3. List a deficiency in SYNOP code that might affect a forecast. Likewise, list a deficiency in METAR code that might affect a forecast.

4. Are winds reported relative to true north or magnetic north?

5. Why do home weather stations generally show lighter, more erratic winds than a National Weather Service station?

6. Why do forecasters choose altimeter setting over sea level pressure in mesoanalysis?

7. If the sky condition is 50% covered by stratocumulus at 4000 feet and the remaining clear sky is half covered by cirrus at 25000 feet, the METAR report will show SCT 040 BKN 250. Why is the cirrus considered broken if it only covers 25% of the visible sky?

8. What has been the effect of ASOS deployment on cloud amount, cloud type, and cloud height measurements, compared to the 1960s?

9. If all of the radiosonde stations became nonoperational due to a bizarre hydrogen or helium supply shortage, what are some alternate sources that could be used for upper air analysis?

10. Which elements are available from profiler data, and which are not?

Figure 2-6. Another federal Form 10, this one for Clark Air Base, Philippines in 1982. This shows the obsolete cloud code format that was used on METAR observations during that era. *(5 July 1982 / via NOAA/NCDC)*

FEDERAL METEOROLOGICAL FORM 1-10 SURFACE WEATHER OBSERVATIONS (METAR/SPECI FORM FOR MILITARY USE)

LATITUDE: 15°11'N LONGITUDE: 120°33'E STATION ELEVATION: +478 Feet(MSL)
MAG TO TRUE: 0 Deg TIME CONVERSION (LST to GMT): −8 Hrs
DAY(LST): 5 MONTH: JUL YEAR: 1982 STA & STATE OR COUNTRY: CLARK AB LUZON R.P.

TYPE	TIME (GMT)	DIR (true)	SPD (knots)	PREVAILING VIS (meters/miles)	RVR LOCAL	WX LOCAL	OBSTR LONG-LINE	SKY CONDITION	TEMP (°C)	DEW PT (°C)	ALSTG (inches)	REMARKS AND SUPPLEMENTAL CODED DATA	STA PRES (inches)	PRES TEND	OBS INIT
SA	1655	E000	00	9999				2SC035 13AC100 13AC150	24	22	2982	CIG100		8	Y
SP	1742	E000	04	9999	1 A.M.			1STOO1 5SC035 3AC100			2980	CIG035		8	Y
SA	1755	E130	05	4800	2 A.M.	Fg	100R	1FG111 5STOO1 5SC035	24	23	2980	CIG035 TWR VIS E 1/6 OVRL3	29.290	8	X
SP	1809	E100	03	3200		Fg	100R	3FG111 5STOO1 5SC035			2980	CIG035 VIS NE1 TWR VIS E 1/6 OVRL 3			Y
SP	1822	E000	00	4800		RA-FG	GIRA	2FG111 5STOO1 5SC035			2979	CIG035 VIS S1 TWR VIS 2			Y
SA	1836	E000	00	4800		FG-	1DBR	2FG111 5STOO1 5SC035			2980	CIG035			Y
SA	1855	E000	00	3200		FG-	100R	3STOO1 13SC035 1AC100	24	22	2980	CIG100		8	X
SP	1916	E000	00	9999		RA-	GIRA	1STOO1 1SC035 3AC100 2AC150			281	CIG150			Y
SP	1940	E000	00	9999			2IRERA	1STOO1 1SC035 2AC100 2AC150 1CI300			281	CIG150			X
SA	1955	E000	00	9999	4 A.M.		2IRERA	1STOO1 3SC035 1AC100 2AC150 1CI300	23	22	2982	CIG150		8	X
SA	2055	E000	00	9999	5 A.M.			2CI300 1STOO1 2SC035 2AC100 2AC150 2CI300	23	22	281	CIG150	29.300	8	X
SP	2108	E000	00	9999		RA-	GIRA	1STOO1 1SC020 2SC035 1AC100			2982	CIG100			X
SP	2136	E000	00	9999			2IRERA	1STOO1 1SC020 1SC035 1AC100			281	CIG100			X
RS	2158	E000	00	9999	6 A.M.	RA-	GIRA	1STOO1 1SC020 1SC035 1AC100	23	22	2982	CIG140		8	RO
SP	2225	E000	00	9999			2IRERA	2STOO1 1SC020 1SC035 1AC100 5AS140			2984	CIG140			RO
SP	2241	E000	00	9999		RA-	60RA	2STOO1 1SC002 1SC020 1SC035 5AS140			2984	CIG035 CIGmOO2 OVR RWY			RO
SA	2255	E000	00	9999	7 A.M.	RA-	60RA	1AC100 3AS140 2STOO1 1SC002 1SC020 1SC035	24	22	2985	CIG035 CIGmOO2 OVR RWY		8	RO
SA	2355	E000	00	9999	8 A.M.	RA-	6IRA	1AC100 3AS140 2STOO1 1STOO5 1SC020 1SC035 1AC100	26	22	2985	CIG140 2OO15	29.335	8	RO
SA	0016	E000	00	9999			2IRERA	3AS140 1CI300 1STOO1 1STOO5 2SC030 3AS140			2985	CIG140			RO
SA	0055	E000	00	9999	9 A.M.		2IRERA	2CI300 1STOO1 1STOO5 2SC030	26	22	2987	CIG030 RADAT 9515Z		8	RO
RS	0159	E000	00	9999	10 A.M.	RA-	6IRA	1CI300 1STOO5 2SC010 6AS140 1CI300	28	23	2985	CIG140		8	RO
SA	0255	E000	00	9999	11 A.M.	RA-	6IRA	2SC030 4AS140 2CI300	28	22	2984	CIG140	29.325	8	RO
SP	0313	E000	20	9999	12 P.M.		2IRERA	2STOO1 2SC002 1SC020 1SC035			2983	CIG140		8	RO
SA	0355	E000	00	9999	1 P.M.	RA-	6IRA	2SC030 4AS140 2CI300	29	23	2983	CIG140		8	RO
SP	0412	E200	06	9999			2IRERA	2SC030 5AS140 2CI300			2982	CIG030			RO
SP	0449	E000	00	9999	1 P.M.		2IRERA	1SC020 4SC030 2AS140 1CI300	29	23	2981	CIG140		8	DH
SA	0459	E000	00	9999	2 P.M.			1SC020 2SC030 3AS140 2AS140	29	23	2980	CIG140	29.285	8	DH
SA	0459	E000	00	9399	3 P.M.			1SC025 1CU030 1ACC50 5AS150	29	23	2978	CIG150 TCU E		8	DH
SA		E000	00	9922	4 P.M.			1SC025 1CU030 2ACO80 1AS120 5AS150	29	23	2977	CIG120 TCU SE		8	DH

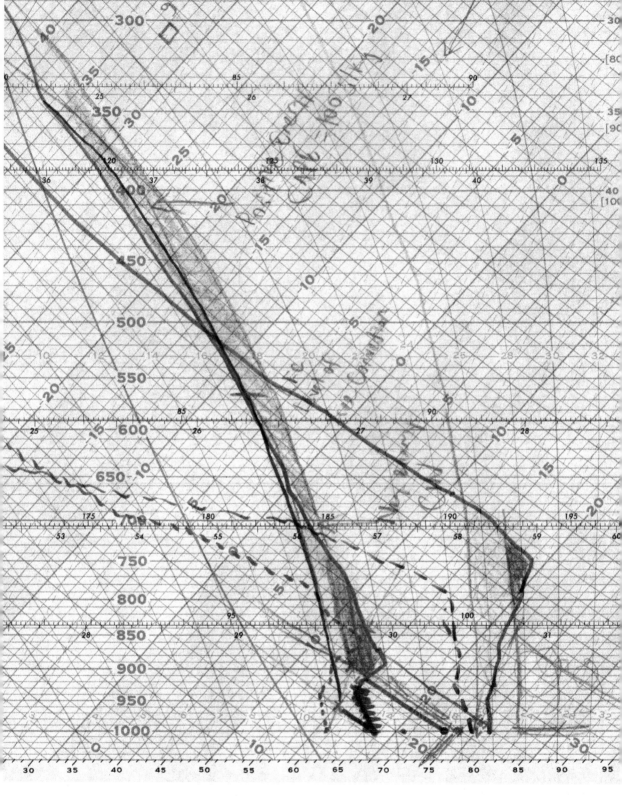

3 THERMODYNAMICS

Weather charts do a good job depicting the big picture, showing where weather systems are and where they're heading. However, they do not depict the actual physics of what's taking place. This is the purpose of the thermodynamic diagram. A moist air mass might have arrived, bringing dewpoints in the 70s, but is it enough for a storm to develop? The temperature is 32 °F; it's raining, and the surface chart shows a low approaching; what's the chance of freezing rain, sleet, or snow? Is the air mass conducive to strong radiational cooling? Thermodynamic tools are designed specifically to help the forecaster solve these questions.

3.1. Phases of matter

Matter exists in three primary states: solid, liquid, and gas. In the solid state the molecules maintain a fixed shape and have a low energy level. In the liquid state, molecules move around but remain cohesive and maintain a constant density. In the gaseous state, molecules are independent of one another and move in any direction, having a high energy level.

Meteorologists deal almost exclusively with phases of water. Solid water is known as ice, liquid water is referred to specifically as a water drop or water droplet, and gaseous water is called water vapor. Vapor should not be confused with forms of visible liquid water like cloud or fog droplets.

These three phases are important because each one is favored at a specific temperature range. The most familiar one is the conversion of liquid to solid ice when the temperature falls below the freezing/melting point. Above that temperature, it will become a liquid. If the substance is heated further to its boiling point, it then tends to change entirely into a gas. Because collisions between molecules are always taking place, a very small fraction of the molecules in liquids or gases are knocked into a high energy state and they escape as a gas. As a result, water does not have to be melted or boiled to produce water vapor.

These changes in energy level are not without consequences. The laws of thermodynamics show that the molecule cannot create or destroy energy to achieve these different states; it has to be transferred somehow. To ascend to a higher energy state, such as from liquid to a gas, the molecule absorbs energy. When you climb out of a pool, some of the liquid on your skin increases its energy state and turns into vapor by absorbing energy from your skin, and as a result you feel cold. Likewise, descending to a lower energy state requires the release of energy by the molecule, and

Changes to higher energy states

Solid to liquid	Melting
Liquid to gas	Evaporation
Solid to gas	Sublimation

Changes to lower energy states

Gas to liquid	Condensation
Liquid to solid	Freezing
Gas to solid	Deposition

Title image
The thermodynamic diagram is actually a workchart. Shown here are several environmental temperature profiles and several parcel lifts. *(Tim Vasquez)*

this imparts heat to its surroundings. To meteorologists this is the release of *latent heat*.

This principle is extremely important because it is responsible for much of the cooling and heating within weather systems. Forecasters must thoroughly understand each of the three states of matter and how heat is transferred when a phase change occurs.

3.2. Adiabatic changes

The term *adiabatic* is used to define a closed system where no thermal energy or mass is exchanged outside the boundaries of that system. In meteorology, we frequently use an imaginary cube of air known as a *parcel*, which is a self-contained system. If we impose an adiabatic process on this parcel, this implies that there is no exchange of heat or mass. The parcel does not absorb heat, radiate heat, or even cast out its water droplets into surrounding parcels. However in some processes, we do in fact specify that an exchange of energy or mass outside the parcel is to take place; such a process is called a *diabatic process*.

Figure 3-1. Processes around mountains can often be described as adiabatic. If air is forced to flow over a mountain, it will cool as it rises and warm as it sinks. This is known as adiabatic cooling and adiabatic warming. In reality, air in contact with a mountain may be affected by things like heating on sunlit slopes and night time radiation into clear sky, so temperature changes are not necessarily a pure adiabatic process. *(Tim Vasquez)*

3.2.1. ADIABATIC COOLING. Anytime a parcel ascends, it rises into levels where the pressure is lower. As a result, the parcel expands. This results in cooling, since the molecules achieve lower energy states due to more spacing and fewer collisions. This cooling is known as *adiabatic cooling*. The rate of cooling with height is given by the *dry adiabatic lapse rate*, which is roughly 10°C per vertical kilometer.

3.2.2. RELEASE OF LATENT HEAT. If a parcel cools to its dewpoint temperature, it has achieved a state of saturation. Any further lifting will cause condensation in the form of water droplets or ice crystals. This phase change by its very nature releases latent heat into the parcel. In turn, this has a very significant effect on the rate of cooling with height. A parcel which is lifted 1 kilometer will be 4 C° warmer from the contribution of latent heat than it would be otherwise. As a result, the overall cooling rate changes from the adiabatic lapse rate of 10 C° km^{-1} to about 6 C° km^{-1}. This is known as the *wet adiabatic lapse rate* or *moist adiabatic lapse rate*. As lifting continues into higher levels of the troposphere, less and less latent heat is released due to dwindling amounts of moisture within the parcel, and the actual wet adiabatic lapse rate value tends to become much larger.

3.2.3. SUBSIDENCE. Now we will look at the sinking of a parcel. When a parcel sinks, or subsides, the pressure rises and molecules are packed closer together. The rate of warming occurs at the dry adiabatic lapse rate, which again is 10°C km^{-1}. So if a parcel starts at 20°C and rises 10 km, it will cool to roughly –80°C, and if we force it back to the surface, it will return to its original temperature of 20°C.

But what if condensation had occurred at some point and liquid droplets are present in the parcel? Once the parcel sinks and warms, it is no longer in a saturated state. The droplets or ice crystals will evaporate back into the parcel. This is a phase change that absorbs heat from the parcel. The parcel warms at the wet adiabatic lapse rate. It may be much warmer than expected if lifted to 10 km, but it will eventually reach the same temperature once back at the surface that it had started at.

It should be noted that in our thought experiment, there is no mass or heat exchange. We kept all water droplets and ice crystals contained in our parcel. In the real atmosphere, these fall out of the parcel due to gravity, while the latent heat remains with the parcel. This process is called *pseudo-adiabatic*. This means that once a parcel begins sinking, no phase changes occur because the liquids and solids are not there. Therefore the parcel warms entirely at the dry adiabatic lapse rate. It warms at a much quicker rate than it would otherwise. This explains why moist Pacific air flowing across the Sierra Nevada range might produce a warm chinook wind on the lee side of the mountains.

Since it is hard to assess exactly how much water will leave the parcel in a given situation, pseudo-adiabatic lapse rates are actually approximations. However, they are fairly close, and in a real atmosphere the error is less than about 1%.

From 120 years in the past
"When air rises from below to above — as over the equator — it commonly has a squeeze from the cold above, and sends down rain. When air sinks from higher to lower levels, it can not only hold all the moisture it held above, but can take in more. So the rising air of a cyclone would naturally cause wet weather; and the sinking air of an anti-cyclone would suck up floating mists, clear the sky, and as a general rule cause sunshine."

THE OCEAN OF AIR, 1890
Agnes Giberne

Figure 3-2.   A radiosonde launch by Shannon Key, a former National Severe Storms Laboratory research assistant and now the author's wife, in Ajo, Arizona during a field experiment. The data acquired from weather balloons yields soundings and hodographs, and are frequently ingested as input into computer model forecasts.

3.2.4. ENVIRONMENTAL LAPSE RATE. So far we have focused only on the temperature within the parcel itself. It's important to point out that the atmosphere has its own distribution of temperatures at different heights, which can be completely arbitrary. The temperature change of the actual atmosphere with height is known as the *environmental lapse rate*. The environmental lapse rates are a reflection of *observed values* of the current state of the atmospheric column. This is a common area of confusion for new forecasters. When we consider the relationship of dry adiabats and moist adiabats to the environmental lapse rate, this yields information about buoyancy and stability.

3.3. Stability

Picture a hot workshop where heat has risen to the top near the ceiling and cool air has settled at the bottom. This is the environment. Now picture a toy balloon. This is our parcel. Let's take our parcel on the floor and force it to the ceiling. Will it float? While sitting on the floor the balloon has taken on the temperature characteristics of the air at the floor, so it is cool. When brought to the ceiling where the surroundings are very warm, the balloon air is relatively dense and it sinks back to the floor. What if another balloon has been idling near the ceiling and is pushed to the floor? The air inside that balloon is already warm. When it reaches the floor it is warmer than the air around it. It is buoyant and returns to the ceiling. So the environment in this workshop is considered *stable*, because when air is displaced vertically in either direction, it returns to its original level.

This is also true in the atmosphere. If a layer has a low environmental lapse rate, that is, the environment cools at a slow rate with height or warms with height, it is said to be stable. Since parcels rise along the dry and warm adiabats, this means that the environmental lapse rate must be less than either adiabatic lapse rate in order to be considered stable. In other words, an environment that cools by less than roughly 6 C° km⁻¹ is stable.

3.4. Instability

Let's return to our workshop and assume that an air conditioner is installed near the ceiling, providing cool air, while a heating mat covers the floor. If our balloon starts at the floor and we tap it upward, it drifts upward slowly, and rises into slightly cooler air. Since the balloon is relatively warm, it continues rising. As it rises into progressively cooler air, it rises at a faster rate until it hits the ceiling. Likewise, a balloon starting at the ceiling is cool, and will descend slowly, but as it encounters progressively warmer air during its descent, it will accelerate downward until it hits the floor because it is relatively cold at any given level. This state is called instability.

The principle is the same in the atmosphere. A layer with a high environmental lapse rate, cooling rapidly with height, is unstable. If we measure this atmosphere, we will find that the environmental lapse rate exceeds the adiabatic lapse rate.

As was outlined earlier, a parcel cools adiabatically at one of two rates, the dry or the wet rate. This means that instability is divided into two types: absolute instability, in which a parcel can be buoyant if dry or wet, and conditional instability, in which a parcel can be buoyant only if it is wet.

Conditional instability simply requires an environmental lapse rate less than the dry adiabatic lapse rate (~10 C° km⁻¹) but more than the wet adiabatic lapse rate (~6 C° km⁻¹). Absolute instability involves an environmental lapse rate that exceeds both the dry and wet lapse rates (i.e., ~10 C° km⁻¹ or more).

In practice, absolute instability is not normal in the atmosphere because when it exists, the layer itself spontaneously overturns to redistribute the heat to a more stable configuration. Take for example the unstable workshop air example above. If we checked it with a thermometer we'd probably find an environmental lapse rate within the workshop of 2 C° m⁻¹, which translates to 2000 C° km⁻¹! This vastly exceeds the adiabatic lapse rate. Not surprisingly, if we turned off the appliances we'd find that the warm air near the floor has risen to the ceiling all by itself, while the cool air has sunk. The workshop air has overturned itself and mixed to a more homogenous temperature. Absolute instability only occurs at the interface between the atmosphere and a warm surface, and should not be found on meteorological soundings.

3.5. Soundings

Now that we've reviewed the principles of thermodynamics, it's time to take a look at our tools. Meteorologists rely heavily on a special diagram called a "sounding". This is simply a graph of properties of the atmosphere such as temperature and dewpoint with respect to altitude (z).

The favored type of graph in the United States, however, is the skew T log p diagram (known simply as a "skew-T"). It was developed in 1947 by meteorologist and physicist Nicolai Herlofson and was quickly embraced by the U.S Air Force,

Absolutely unstable layer
* Slope: The lapse rate is steeper than the dry adiabat.
* Lapse rate: Greater than about 10 C deg per km.
* When a parcel is forced upward: It cools slower than the environment, so it becomes more buoyant and continues accelerating upward.
* When a parcel is forced downward: It warms at a slower rate than the environment, so it becomes heavy and continues accelerating downward.

Conditionally unstable layer
* Slope: The lapse rate is steeper than the wet adiabat but not as steep as the dry adiabat.
* Lapse rate: Greater than about 6 C deg per km but less than about 10 C deg per km.
* When a parcel is forced upward: If it is dry, it cools at faster rate than the environment, so it becomes heavy and returns to its original level. If it is saturated, it cools at a slower rate than the environment, so it becomes more buoyant and accelerates upward.
* When a parcel is forced downward: It warms at a faster rate than the environment, so it becomes buoyant and returns to its original level.

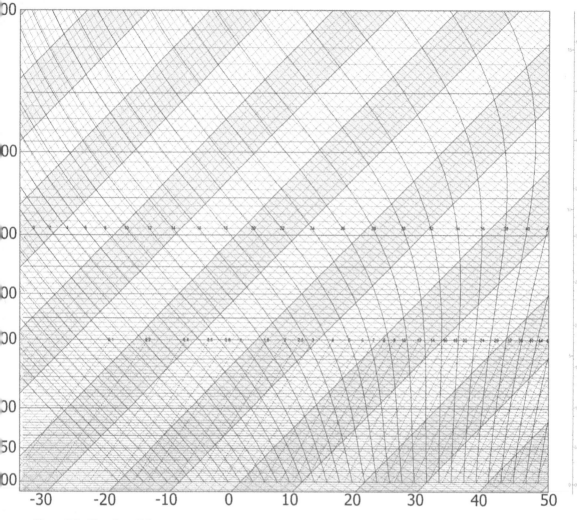

Figure 3-3. The skew T log p diagram. This is the backbone of thermodynamic work in meteorology. *(Tim Vasquez)*

gradually gaining acceptance in the National Weather Service. It's by far the most common chart used in the United States and Canada, and is what will be used throughout this book. It is composed of the following items.

3.5.1. PRESSURE LINES. Pressure (height) lines run left-right on the chart. To plot the position of a parcel of air, we put a dot higher on the chart if it is at a higher height. The lines are marked in millibars, but many diagrams also have a height scale to the far right which indicates feet and meters.

3.5.2. TEMPERATURE LINES. Temperature lines are straight solid lines running from lower left to upper right on the chart. The lines are marked in degrees Celsius. They are used to determine the temperature or dewpoint of a parcel at a given height.

3.5.3. MIXING RATIO LINES. Mixing ratio lines are straight dashed lines running from lower left to upper right on the chart. They

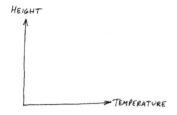

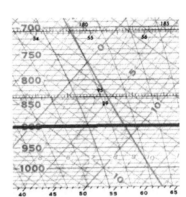

Figure 3-4a. Soundings are not confusing when you keep in mind that they simply allow us to plot a weather reading on a graph of height versus temperature. If the lower atmosphere has a high temperature, you plot a dot at the bottom right. If the upper atmosphere has a low temperature, you plot a dot at the top left. The dots can be connected to show a temperature profile.

Figure 3-4b. Pressure (height) lines run parallel to the thick line shown.

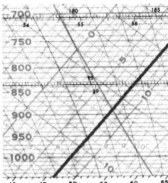

Figure 3-4c. Temperature lines run parallel to the thick line shown. Together, the pressure (height) and temperature line help locate the temperature of a parcel at a given height. Everything else is secondary.

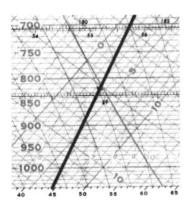

Figure 3-4d. Mixing ratio lines run parallel to the thick line shown. They help identify the mixing ratio or the saturation mixing ratio of an air parcel.

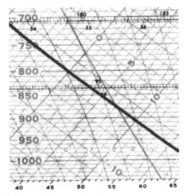

Figure 3-4e. Dry adiabat lines run parallel to the thick line shown. When a parcel rises or sinks without any phase change, it can only do so along a dry adiabat. More about this is covered later in the text.

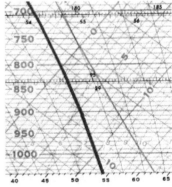

Figure 3-4f. Wet adiabat lines run parallel to the thick line shown. When a parcel rises and it is saturated (condensing), it can only rise along a wet adiabat. Parcels only sink along dry adiabats. More about this is covered later in the text.

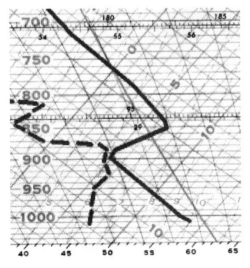

Figure 3-5. An example of a sounding. The temperature profile is indicated by the solid line on the right. It shows a temperature of 14 deg C at the surface, 5 deg C at 900 mb, 7 deg C at 850 mb, and so forth. The dewpoint (moisture) profile is indicated by the dashed line on the left. It shows a dewpoint of 7 deg C at the surface, 5 deg C at 900 mb, -3 deg C at 850 mb, and so forth.

are marked in g/kg. These lines are used to determine the mixing ratio or saturation mixing ratio of a parcel at a given height.

3.5.4. DRY ADIABAT LINES. These are slightly curved lines running from lower right to upper left. They are marked in units of potential temperature in degrees Kelvin. A parcel always follows this line if it is dry (not saturated) and is rising or sinking.

Figure 3-6. An actual sounding. This shows that the environment contains a moist, mild zone of air in the lowest 5000 ft of the atmosphere, capped by a relatively warmer layer just above. *(2 April 2011 / 1200 UTC)*

3.5.5. WET ADIABAT LINES. These are curved lines running from lower right to upper left. They are marked in units of potential temperature in degrees Celsius. A parcel always follows this line if it is saturated (condensing) and is rising. Sinking parcels only follow the dry adiabat lines.

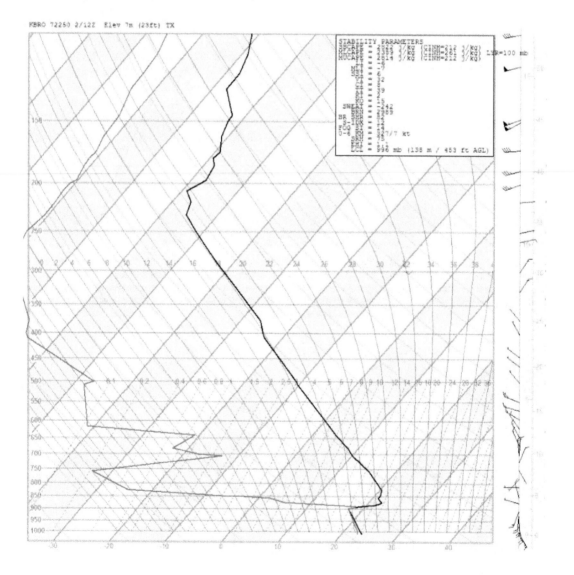

3.5.6. PLOTTING THE SKEW-T. To use the skew-T, a forecaster must first plot the *environmental data*, obtained from actual radiosonde observations. First, temperature is plotted from the surface to the top of the chart. This is done in solid marking, preferably in red or black. Then dewpoint temperature is plotted in a similar manner using a dashed green or black line. It's almost impossible to get the two lines confused because the temperature line will always lie to the right of the dewpoint line, since dewpoint cannot exceed temperature.

Forecasters may also plot wind data. It is not possible to do computations on wind data using the skew-T diagram, but the information it provides can help identify the height of jet streams and suggest whether a lot of shear and helicity might be present or not.

3.6. Sounding interpretation

Once the environmental data is plotted, the sounding is ready for interpretation and forecast use.

3.6.1. ASSESSING INSTABILITY. The sounding makes it easy to determine whether individual layers are unstable or not. All you have to do is look at the plotted temperature and check how the slope matches up with the dry adiabat and wet adiabat lines. If the temperature plot leans far to the left with height, this indicates the layer cools rapidly with height. If its slope exceeds that of both adiabat lines, then the layer is absolutely unstable (which of course is very rare), but if it falls between the two, the layer is conditionally unstable. If it leans to the right of the adiabat lines with increasing altitude, then the layer is stable.

3.6.2. INVERSIONS. Layers that are highly stable, in which the temperature is the same with height or increases (leaning rightward with height relative to the temperature lines) are known as *inversions*. These are nothing but layers of warm air above cool air, and the stability prevents air from moving between the layers.

There are three key types of inversions encountered in operational meteorology: frontal, radiational, and subsidence. The *frontal inversion* layer usually shows an increase in humidity with height (i.e., the temperature and dewpoint lines converge with height), and this is a reflection of the cloud layers present within the transition zone. The radiational inversion is caused by cooling from below, making it very common on clear nights, and is recognized by the decrease in humidity in the inversion layer

Figure 3-7. Parcel lift. The environmental sounding is given by the lines T - T' (temperature) and Td - Td' (dewpoint). Starting with this information, a parcel is lifted from the surface temperature along the dry adiabat until it reaches the mixing ratio line corresponding with the surface dewpoint. This is the LCL. From that point it rises along the moist adiabat. the result is a lifted parcel temperature profile X - X'. The lifted parcel dewpoint profile is given by the line segment Td - LCL - X'. This parcel lift method is the surface-based (SB) method. Forecasters may instead choose to use the portion of the sounding that yields the most buoyant parcel (most unstable, MU, method) or average out the low level air to use for a parcel (mixed layer, ML, method).

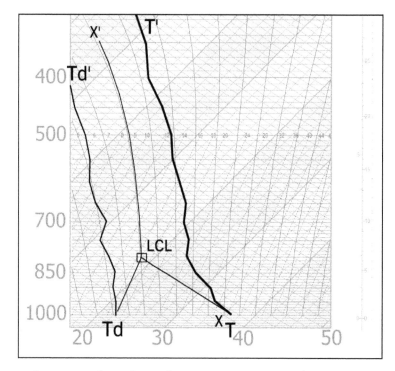

and appears only at the surface or, in quiescent weather regimes, at the top of thick cloud layers. The *subsidence inversion* is caused by warming from above, typically from downward motion. It is almost identical in appearance on the diagram to a radiational inversion but the layer is more likely to exist aloft rather than at the surface. The elevated mixed layer (covered later) tends to produce subsidence inversion signatures.

3.6.3. Elevated Mixed Layer (EML). Warm air originating from higher terrain may advect with the prevailing flow and spread over the top of existing air masses elsewhere. This effect frequently occurs in the central and eastern United States. The Rocky Mountain region has a mean elevation of 6800 ft in Colorado, 5700 ft in Utah, and 4100 ft in Arizona, and this region is very warm in terms of temperature normalized to sea level. As a result, strong westerlies blowing across the southwest United States will blow this air over existing air masses east of the Rockies, producing a warm-over-cold situation even if a warm tropical air mass is at the surface. This warm layer aloft, once known as a "Superior air mass" in the 1940s, is called the *elevated mixed layer (EML)*, and it forms a *capping inversion*, or simply a *cap*, on the sounding, bounded on the western periphery by a surface dryline. The EML shows subsidence inversion characteristics, though advection is largely responsible for the inversion.

The EML and cap are significant because they tend to suppress deep convection until additional amounts of moisture and heat become available. This results in much stronger instability than would occur otherwise, and makes severe thunderstorms more likely. This situation forms the signature for the "Miller Type I" or "loaded gun" sounding. The EML and cap are most common in the United States and adjacent parts of Canada and Mexico, but any regions downstream from a warm plateau area are susceptible. The pattern has been observed in India, Bangladesh, China, Argentina, and Central Europe, and may occur on a smaller, more local scale in Australia and South Africa.

3.6.3. PARCEL LIFT CONSTRUCTION. Parcel lifts are done on the skew-T by forecasters to determine whether the atmosphere might support deep convection — the growth of a cumulus or cumulonimbus cloud through the troposphere in a conditionally unstable atmosphere.

The simplest way of doing a parcel lift is to lift a parcel from the surface. The surface temperature point and surface dewpoint point are located at points defined by the very bottom of the environmental temperature and dewpoint trace. The parcel's temperature is drawn starting at the surface temperature point, following the dry adiabat until it intersects the same mixing ratio line that intersects the point representing surface dewpoint. This height (P, or z) is called the lifted condensation level (LCL), and is where the parcel becomes saturated and cloud material and water droplets form. From there, it proceeds upward along the moist adiabat to the top of the diagram. This method is called the *surface based (SB) parcel method*.

Parcels which form part of a cumulus cloud are never composed entirely of surface air. Some of the subcloud air also gets entrained into the cumulus cloud. This calls for the forecaster to try lifting an average of the properties near the ground. This is called the *mixed layer (ML) parcel method*. Figure 3-8 provides guidance on how to determine an average starting parcel temperature and dewpoint. From there, the parcel is lifted just the same as the surface based method.

In situations where there is elevated moisture, the parcel that provides the greatest buoyancy might not come from the ground but from some level aloft. A parcel that has the greatest buoyancy is the one that has a lifted parcel temperature line that is as far to

1927: The elevated mixed layer

Mid-level temperature inversions appear to be quite common and the lapse rates above these inversions increase very rapidly, often nearly or quite of adiabatic value. In short, so far as one can infer from these few observations, the atmosphere in the neighborhood of a tornado appears to be unusually stratified, and tending to become unstable at one or more levels.

W. J. HUMPHREYS, 1927
U.S. Weather Bureau

the right as possible. This calls for a variation known as the *most unstable (MU) parcel method.*

First let's go back to the surface based method where a LCL point was created. If we follow this point with our finger along the wet adiabat line back down to the parcel's starting level and read the temperature scale, this gives us the starting wet-bulb temperature. If we continue following this line to the 1000 mb level, this gives us a quantity known as wet bulb potential temperature, or theta-w (θ_w). We can also follow this adiabat to the top of the diagram where it becomes parallel with the dry adiabats, then sink it along the dry adiabat to 1000 mb, which yields equivalent potential temperature, or theta-e (θ_e).

The most unstable level, therefore, is the level that exhibits the highest value of θ_w or θ_e. If that level is lifted, the parcel will join a wet adiabat that is further to the right than any other parcel. The easiest technique is to simply force the computer to plot a profile of environmental wet bulb temperature, which will fall between the temperature and dewpoint line. The level where the environmental wet bulb reaches furthest to the right relative to the wet adiabats is the most unstable level. If wet bulb temperatures cannot be plotted, the forecaster can systematically pick levels on the skew-T and lift parcels to their LCL; the LCL point that is positioned furthest to the right relative to the moist adiabat was lifted from the level of most instability. The most unstable method is recommended whenever an elevated moist or warm

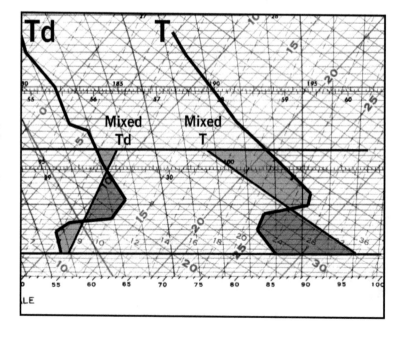

Figure 3-8. The technique of mixing a layer manually. Using the sounding, the forecaster selects a dry adiabat (or an imaginary line parallel with neighboring adiabats) which bisects the temperature trace so that an equal geometric area exists on each side of the adiabat. Likewise, for the dewpoint trace, a mixing ratio line is selected which bisects it to form equal areas. In this case, two areas exist on one side of the line, which can be simply combined before evaluating.

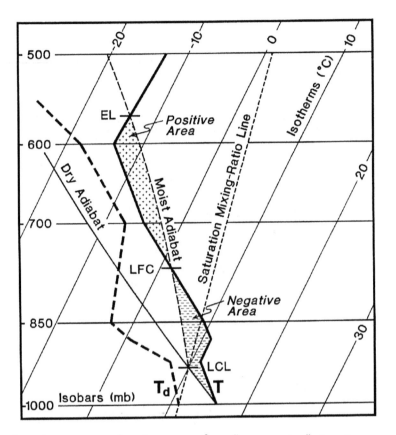

Figure 3-9. Summary of sounding features showing the interrelation of a lifted parcel with its environment. In this example we assume a surface temperature of 15 deg C and a surface dewpoint of 11 deg C. The environmental temperature profile for this day is drawn as a thick solid line (T), and the environmental dewpoint profile is drawn in thick dashed line (T$_d$). We lift a parcel from the surface, noting its starting mixing ratio. The parcel's temperature always follows the dry adiabat until it meets this mixing ratio line. At this level, condensation occurs, and we consider this level to be the lifted condensation level (LCL). From that point, it follows the moist adiabat upward. This lifted temperature profile is compared against the environmental dewpoint profile. At levels where the lifted parcel is cooler than the environment, a negative energy area is indicated (the parcel is cooler than the surrounding air, thus it gains a sinking tendency). At levels where the lifted parcel is warmer than the environment, a positive energy area is indicated (the parcel is warmer than the surrounding air, thus it gains buoyant tendencies). Wherever the lifted parcel rises into an area of large positive energy, this level is called the level of free convection (LFC). The parcel rises until the surrounding temperature becomes warmer (typically in the stratosphere); the level at which this crossover occurs is called the equilibrium level (EL). The parcel will usually continue rising through its own momentum, and will sink back under the EL with time.

layer is present, such as in a warm front "overrunning" situation, and the surface layers are too cool or dry to support convection.

The most unstable method lifts only a single level, however, which again is unrealistic in a weather situation. When elevated moisture is present, meteorologists should turn to an elevated mixed layer technique. This is identical to the ML method, except that instead of binding the mixed layer to the surface, an appropriate layer aloft is selected. This method is recommended for elevated convection and warm frontal situations. If the forecaster does not understand this method, the situation then calls for the most unstable method.

3.7. Instability quantification

Since the 1950s, forecasters have attempted to quantify instability and other parameters to provide a consistent methodology for anticipating severe weather. An algorithm or calculation which summarizes the stability or instability in some way is known as a *stability index*. Over the years, many have even been optimized to even provide forecast guidance and risk probabilities for severe weather. However, it is telling that very few of these indices have endured. With the recent paradigm shift to ingredients-based forecasting they are now used very sparingly.

One of the drawbacks of many stability indices is that they are rigid in nature; many use only two or three key levels and this may cause the forecaster to completely overlook a significant feature, such as an inversion or a layer of moisture. Furthermore, the sounding must be modified to account for conditions at convection time before applying the methods; an index based on a 1200 UTC sounding is often unrepresentative of storms that develop at 2100 UTC. This is an extremely common pitfall when using stability indices from Internet products.

In general, the favored stability index is CAPE, which is an expression of buoyancy that uses the entire environmental sounding. This is measured in joules per kilogram, and is directly related to updraft strength. A value of 500 j kg⁻¹ or more is usually associated with thunderstorms and 2000 j kg⁻¹ a common value with severe weather in the U.S. However even CAPE requires the use of a correctly-selected parcel and an accurately modified environmental temperature profile.

3.8. Potential instability

Potential instability is a different type of instability altogether. Rather than lifting a parcel, we lift an entire layer (i.e. every parcel in the layer). This is frequently what occurs with deep sources of lift such as upper level divergence. The result is that saturated

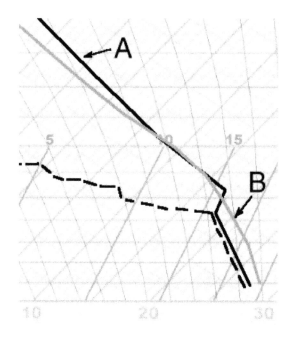

Figure 3-10. Example of potential instability (see text). The entire layer represented by the temperature trace (dark solid line, A) is lifted by 100 mb. The result is the solid gray line (B). It can be seen that the lower layers have warmed (due to release of latent heat), while the upper layers have cooled (due to adiabatic cooling). The result is a temperature profile that has a steeper lapse rate and is trending towards becoming absolutely unstable. Note that the mid-level inversion has been eroded by the lift.

levels where latent heat is released tend to cool slowly, while dry levels tend to cool rapidly. If the saturation is in the low levels rather than the upper levels, this steepens the environmental lapse rate, making the atmosphere more unstable. Such a situation frequently occurs when major weather systems approach a moist tropical air mass.

3.9. Symmetric instability

Instability is any process in which an initial displacement results in an acceleration in that direction. Convective instability describes a vertically-oriented stability. With symmetric instability, a parcel is considered stable in the vertical and horizontal directions, and will return to its original position if displaced. However if it is moved in a slantwise direction, it is unstable.

Symmetric instability is related to the configuration of the geostrophic absolute momentum (M_g) surfaces against the equivalent potential temperature (θ_e) surfaces. A parcel always tries to conserve M_g and will "adhere" to those surfaces. Normally a cross section will show that the M_g isopleths are vertical and θ_e isopleths are horizontal, so that when a parcel rises, the θ_e increases with height. If the M_g isopleths are somewhat horizontal and the θ_e isopleths are vertical, this will cause a rising parcel

Figure 3-11. Simplified diagram of symmetric instability regimes as viewed on a cross section. Solid lines are geostrophic momentum, while dashed lines are saturated equivalent potential temperature. A conditional symmetric instability (CSI) area appears where the slope of momentum lines is not as vertical as theta-e lines. Where they are parallel with one another, a weak symmetric stability (WSS) area is suggested. Note that where the potential temperature isopleths fold over so that values fall with height instead of increase, regardless of momentum isopleth orientation, a convective instability regime (CI) is indicated. Areas outside all of these regions are stable.

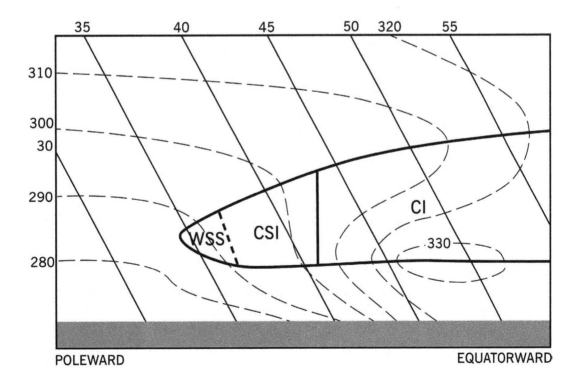

to encounter lower θ_e values with height, causing an acceleration. However, that said, if the θ_e surfaces are vertical enough to "fold over" to produce a lowering of θ_e in the vertical, this actually causes a convectively unstable layer, not a symmetrically unstable layer.

How can the momentum isopleths be tilted out of the vertical to achieve symmetric instability? The simplest expression of absolute momentum is $M_g = v_g - f$, showing that it equals the geostrophic wind speed minus Coriolis effect. If the winds increase sharply with height, M_g will change sharply with height, producing drastically higher values of M_g with height and causing the isopleths to become more horizontal. So strong bulk shear is important in symmetric instability environments.

3.9.1. SYNOPTIC PATTERNS.

The patterns that favor symmetric instability are a saturated atmosphere, an environmental temperature profile that is close to moist adiabatic, strong speed shear in the vertical, and a source of lift. This commonly occurs in areas of frontogenesis and near anticyclonically curved entrance regions of jets, including the subtropical jet. If equivalent potential vorticity (EPV) fields are available, negative values or values less than 0.25 units indicate an area of possible CSI.

3.9.2. SYMMETRIC INSTABILITY IDENTIFICATION.

Forecasters will use a cross section perpendicular to the thermal wind (thickness isopleths) which shows momentum (M_g) and equivalent potential temperature (θ_e). Ideally, saturated equivalent potential temperature (θ_{es}) should be used, but it is harder to obtain and θ_e will suffice as long as the relative humidity is very high. In short, CSI is found where the momentum isopleths are more horizontal than the theta-e isopleths. The theta-e values should always increase with height, but in layers where they don't, convective instability is the dominant process. There are limitations inherent in analyzing for CSI due to the poor resolution of radiosonde network and changes in geostrophic quantities such as M_g in ageostrophic flow.

3.9.3. SLANTWISE CONVECTION.

Given sufficient moisture and a source of lift, CSI often results in slantwise convection. This appears as mesoscale fields of weak convective precipitation oriented in bands which are parallel to the thermal wind. The precipitation fields can be extensive with the heavier bands "embedded" in the precipitation area. However, since this is not a convectively unstable regime, the vertical motion is weak, and thunder and lightning usually does not occur. However precipitation can be very heavy. The cloud form is best characterized as nimbostratus.

Chapter Three
REVIEW QUESTIONS

1. When matter changes from solid to a gas, is this a change to a higher energy state or a lower one? Does this absorb heat or release heat into the environment?

2. Why does a parcel cool when it rises?

3. Does the release of latent heat augment or dampen ascent?

4. Describe the difference between an adiabatic process and a pseudo-adiabatic process.

5. Why are large areas of absolute instability unlikely to be seen on soundings? If they did occur, what would you see that indicates an absolutely unstable layer?

6. What might an inversion look like on a sounding?

7. When a parcel rises along the wet adiabat instead of the dry adiabat, why has its rate of cooling changed?

8. Why is it often not a good idea to lift parcels using the surface based method?

9. The stability indices from the 1200 (7 a.m. CDT) sounding for Omaha showed no instability, yet severe thunderstorms developed later in the day. There was no significant change in the air mass. What might explain this?

10. Explain a weather pattern that might be associated with conditional symmetric instability.

```
unch Time (y,m,d,h,m,s):          1999, 07, 20, 23:55:10
Type/ID/Sensor ID/Tx Freq:        VAISALA RS80-15LH 0, 0,
ocessor/Met Smoothing:            NCAR RS80 PROCESSOR, 10
Type/Processor/Smoothing:         LORAN-C, ANI-7000, 60 SE
aunch Met Obs Source:             MANUAL ENTRY
m Operator/Comments:              SKEY,
```

```
OH III  THIS DATA SET IS UNEDITED AND HAS NOT BEEN REV
e   Press  Temp  Dewpt   RH    Uwind   Vwind   Wspd   Dir
c    mb     C      C     %      m/s     m/s     m/s    deg

0.0  938.2  33.2  10.8  25.5   -2.7    -.2     2.7   86.8
0.0  931.6  32.7  10.9  26.4   -2.7     0.0    2.7   89.4
0.0  926.3  32.1  11.1  27.5   -2.0     .5     2.1  104.8
70.0  920.7  31.6  10.9  28.1   -1.8     .4     1.9  102.7
80.0  917.5  31.2  10.8  28.4   -1.7    -.1     1.7   87.4
90.0  914.2  30.9  10.8  29.0   -1.8    -1.0    2.1   60.9
100.0  910.6  30.5  10.8  29.7   -1.6    -1.4    2.2   49.0
110.0  907.2  30.2  10.7  29.9   -1.6    -1.9    2.5   40.0
120.0  904.0  29.9  10.7  30.4   -1.1    -2.2    2.5   27.3
130.0  900.8  29.6  10.6  30.8    -.5    -1.9    2.0   15.7
140.0  896.3  29.2  10.5  31.3    -.1    -1.2    1.2    4.4
150.0  892.8  28.9  10.5  31.9     .1    -.4      .4  342.3
```

```
1  10 Sec Data   2  Raw Met Data   3  Navaid Status   4  Navaid He
6                 7                 8                  9  PRIMARY M
```

DELL Latitude LM

4   UPPER AIR ANALYSIS

ooking at television weather, it would seem that the proper starting point for a forecast is the surface map with its smorgasbord of fronts, highs, and lows. But like a traffic helicopter that gets the big picture from above, a forecaster should begin with a *top down* approach using the upper air charts. These charts serve as a "master key" that helps to break the weather situation down to its most simple constituents. They also suggest where many of the fronts, weather systems, and air masses might be found on the surface chart.

Though the upper level charts provide a good starting point for a forecaster, there is really no wrong sequence for doing an analysis. After all, the low-level weather and upper-level weather are interdependent parts of a large atmospheric system. A change at one level often influences all other levels. Consequently, the weather forecast is not entirely driven by the upper-level charts. Low-level patterns may very well be the key to the forecast on a particular day. However, the upper-air analysis is not heavily affected by friction, heating, evaporation, and weather phenomena, and the higher one ascends, the more the patterns tend to simplify. This means that a top-down approach allows the forecaster to start with the most basic, digestible patterns at hand and gradually work into the finer details of the system.

Where does the data come from for an upper-air chart? Soundings are taken twice a day at hundreds of stations around the world. The sensor, about the size of a milk carton, is launched on a weather balloon. During its 45-minute ascent, the sensor sends a telemetry signal back to the station containing pressure, temperature, and humidity data. This data is collected by radiosonde processing software and is formatted into an internationally-standardized radiosonde report format known as TEMP (WMO FM-35 or TTAA) format. It is then distributed through weather networks worldwide. Although this numerical data can be decoded by hand and directly plotted on charts by skilled forecasters, many software programs and Internet sites relieve the forecaster of this burden. The values from the TEMP format are also used to produce upper air charts (covered earlier).

Upper-level charts are easy to construct over places like North America and Europe. However, radiosonde data is rarely available over ocean basins and third-world countries. The simple construction of an upper-air chart for places like Africa usually results in a blank map. Therefore, forecast agencies are heavily dependent on two additional sources for a complete upper-air chart: AMDAR reports of wind and temperature collected from commercial aircraft and transmitted automatically by the onboard flight management systems, and satellite-derived wind

Radiosondes

Much interest has been shown recently in developing meteorographs which are capable of transmitting by radio a record of conditions which they encounter as they ascend into the atmosphere attached to small free balloons. A number of such types have been developed within the last few years, some of which have yielded very satisfactory results. This method of obtaining upper air information promises to supplant the use of airplanes in the very near future.

GEORGE F. TAYLOR
"Aeronautical Meteorology", 1938

Title image
Real-time feed from a weather balloon, photographed by author while participating in the NSSL SWAMP project in Arizona, *(Tim Vasquez)*

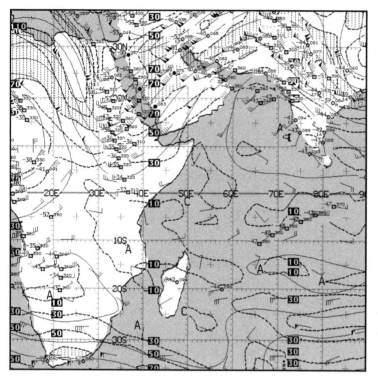

Figure 4-1. Actual upper-level observations of Africa and the West Indian Ocean region on March 15, 2011. This chart demonstrates the scarcity of radiosonde observations (circle plots) in some areas, existing only in South Africa, Madagascar, and the industrialized countries of the Middle East. The data is heavily supplemented by AMDAR aircraft reports (square plots) from commercial aircraft and satellite-derived wind observations (plots with no symbol at the station location). This illustrates that feeding a forecast model an accurate initial state of the atmosphere is a very complicated undertaking.

observations, produced by specialists who monitor conditions at regional forecast centers.

4.1. Constant pressure charts

Starting in the 1950s weather agencies began embracing the constant pressure chart. On this type of map, we see conditions at a constant pressure, in other words, at an altitude where the instrument packages on weather balloons fell to a specific pressure value while rising from sea-level to the partial vacuum of the stratosphere. For instance, the most popular type of constant pressure chart is 500 mb. In polar regions, this is found at about 5 km MSL and in tropical regions at about 6 km MSL.

On a constant pressure chart, pressure patterns are depicted not by isobars but by contours (isohypses), which connect areas of equal geopotential height. For example, on a 500 mb chart we would see low heights in the polar regions, where the surface height averages about 5 km, and high heights in the tropical regions. Areas of high heights are in fact equivalent to high pressure areas, and vice versa.

Specific advice on plotting and coloring constant pressure charts are listed in the appendix.

4.2. Long waves

High in the atmosphere, the winds and heights are organized into very large waves measuring thousands of miles in size. These are called long waves, or "Rossby waves". They are usually masked by smaller-scale waves embedded within them. On a typical day there are four or five long waves encircling the globe in the temperate latitudes of both the northern and the southern hemisphere.

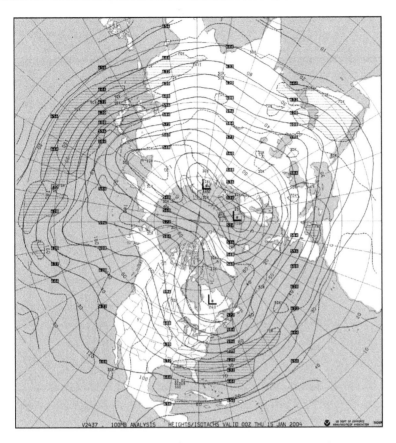

Figure 4-2. **The 100 mb level** is located well up in the stratosphere, but this chart gives a sense of how the circulation at that level is made up almost entirely of very broad waves. This chart shows three long waves around the hemisphere, with a trough in the western Atlantic, another in eastern Eurasia, and another in the west Pacific. These are long waves. The diagram at left shows the nature of upper-level waves.

Long waves take up a scale the size of a continent. They are about 50 to 120 degrees of longitude in wavelength, have an amplitude of 1500 to 1800 miles, and are best seen at 300 and 200 millibars (above 30,000 feet). Most of the irregularities seen in the lower layers have dampened out at these heights, and you'll see broad, large-scale ridges and troughs.

There are usually 4 or 5 long waves around the hemisphere, but there can be as many as 7 or as few as 3. Their amplitude is 1500 to 1800 miles, and they have a wavelength of 50 to 120 degrees of longitude. They move eastward at up to 15 knots or may be stationary or retrogress at up to 3 knots.

Long wave ridges are generally associated with warm weather, with the area downstream typically experiences fair weather.. Long wave troughs are associated with cold temperatures, and the area downstream typically gets unsettled or stormy weather, especially if the long wave trough is deep enough to dip into subtropical moisture sources. This often occurs in the southern U.S. in the wintertime.

When long waves are few, they will have larger wavelengths and will move slower. As a general rule, if there are only 3

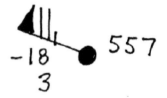

Figure 4-3. **Upper-level plot model**. This example is extracted from a 500 mb chart. The temperature is -18 deg C, and dewpoint depression is 3 deg C (therefore the dewpoint is -21 deg C). The height of the constant pressure surface is 5570 meters (the translation of these digits varies depending on the pressure surface being used; see text). The circle is shaded to indicate a dewpoint depression of less than 5 deg C (moist air). The winds are out of the west-northwest at 75 kts.

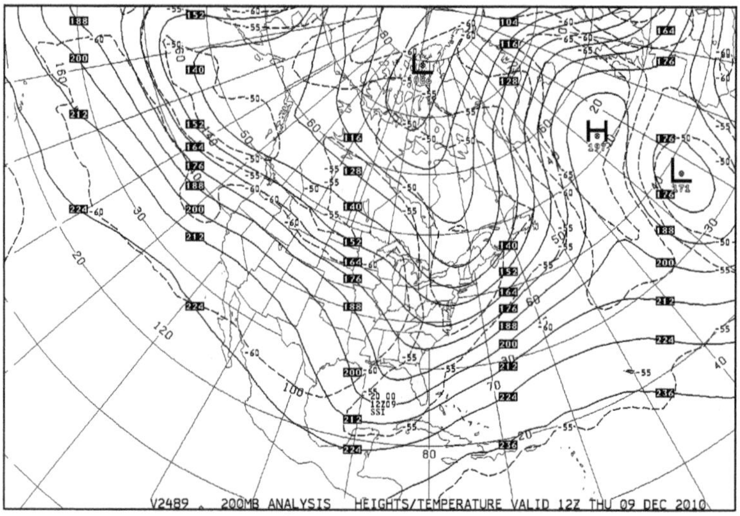

Figure 4-4a. High-zonal pattern.
Note how the flow is predominantly west-to-east, with only minor variations. This pattern is generally unstable in nature and the upper-level flow soon tends to break up into a series of waves. This pattern is called simply "zonal".

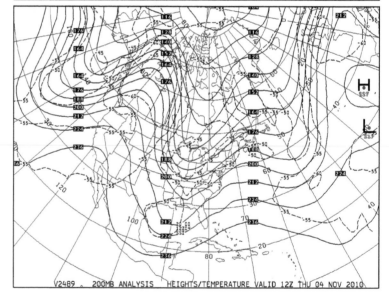

Figure 4-4b. Low-zonal pattern.
This generally represents a breakup and cellular fragmentation of the upper-level flow. Note how the flow is almost direct southerly over the Pacific coast, turning sharply to a northwest flow over the Great Plains. This pattern is also known as "meridional".

Figure 4-5. Troughs and ridges in the upper troposphere. Since the polar regions contain low pressure (low heights), any southward expansion of contours and flow from the pole defines a trough. The opposite is true for a ridge. If you're thinking that upper-level troughs are associated with cold polar air masses, you are indeed correct.

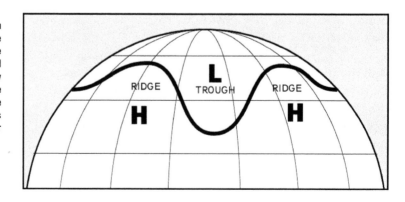

long waves, the pattern around the hemisphere will retrogress. Retrogression is rare but it does occur from time to time. At 4, the pattern will remain stationary or move slowly, and at 5 or 6, the long waves will progress from west to east.

During the winter, a long wave trough pattern tends to cover a large part of continental landmasses while ridges prevail over the open ocean. This is because the continental area has a lower mean temperature, with lower vertical thicknesses that translate to lower heights aloft. The opposite effect occurs during the summer season.

A long wave ridge will build when warm air advection or a major short wave ridge (to be covered shortly) moves into the long wave ridge. A jet streak moving into the back side of a long wave ridge will cause it to build, and if there is a deep trough further downstream, the low may close off. A jet streak moving out of the ridge will cause it to weaken. If a jet streak is moving around a sharp ridge and a northeast-southwest long-wave trough is downstream, the trough will fill and orient from north-to-south. If a jet streak is moving around a sharp ridge and a north-south trough is downstream with a blocking ridge further downstream, the trough will fill.

Long wave troughs will deepen when cold air advection or a short wave trough moves into the long wave. A jet streak moving

Long waves

* Long wave troughs and ridges may be difficult to determine from a single 500 mb chart.

* Eighty percent of the time long wave patterns do not move.

* Short wave troughs deepen at the base of the long wave position.

* Short wave troughs can flatten mean ridges resulting in a zonal chart.

* Surface cyclogenesis takes place east of stationary long wave trough axes. The surface low tracks northeast at this time.

* Short wave troughs tend to be difluent upstream and confluent downstream of the long wave position.

S. BUSINGER,
"The Long Wave Concept", 1987

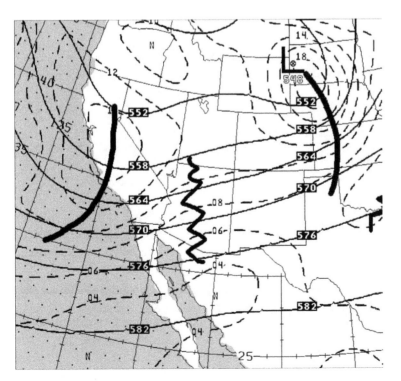

Figure 4-6. Short waves hand-drawn on top of a chart showing 500 mb contours (solid) and absolute vorticity (dashed). Each trough, represented by a thick solid line, represents where the highest level of cyclonic vorticity is along a particular contour, and each ridge, indicated by a zigzag line, is where anticyclonic vorticity is highest. Where vorticity lines cross the contour lines, it forms a webwork of imaginary boxes. The smaller these boxes are, the stronger the vorticity advection.

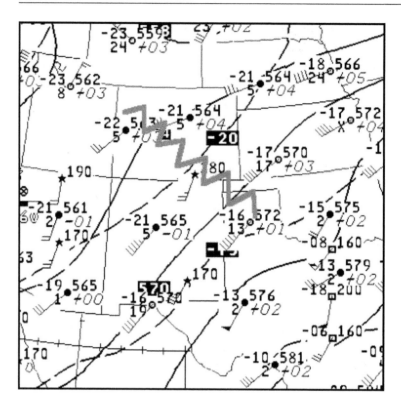

Figure 4-7. A "short wave" does not always mean a short wave trough. Here we see a short wave ridge. It is drawn here with the jagged line and is placed where anticyclonic turning of the winds is present.

into the back side of a long wave trough will deepen it, while a jet streak moving out of it will cause it to fill.

4.3. Short waves

Short waves are small-scale waves associated with temperature advection. They show up best in the mid-troposphere, at about 700 mb, but many of the larger ones may extend up to 500 mb. It is important to remember that short waves include both ridges and troughs.

Short wave features are mesoscale in size, which means that they can be difficult to locate on standard radiosonde charts, which have synoptic-scale spacing. They are embedded in the long-wave pattern. Their wavelength is anywhere from 1 to 40 degrees of longitude (those over 15 degrees are considered "major short waves") and they have an amplitude averaging 100 to 1000 miles. They are small, they often contain intense areas of vertical motion.

Short wave troughs [ridges] are best found by locating a cyclonic [anticyclonic] wind shift in the prevailing flow and/or a thermal trough [ridge] in the upstream direction. Vorticity charts are another tool that can be used to locate short waves; this is discussed in a later section.

Wind intensity

What constitutes a strong wind or a weak wind at various levels? Here is a quick reference guide, which will vary according to location and season. Speeds are listed in knots. 850 mb: Weak

Type	Level (mb) 850	500	250
Weak	<20	<35	<55
Mdt	21-35	36-50	56-85
Strong	>35	>50	>85

Why do short wave troughs appear to "cause" weather? They are not really an object, but a process. Elements like cyclogenesis, lift, and low-level warm air advection are all capable of maintaining the slightly higher heights ahead of the short wave trough. The heights within the short wave trough are lower due to cooling within the column, which may destabilize the atmosphere if differential thermal advection is occurring. All of these factors tend to increase the likelihood of upward vertical motion as the short wave trough approaches.

4.4. Divergence and convergence

Divergence of the wind field is a measure of the rate at which mass is removed from a given volume of air. It results in a decrease of atmospheric mass above a given point, and the surface pressure falls. This is because the equations for force and pressure show that removing mass decreases downward force, and decreasing force decreases pressure. The opposite is true for wind field convergence.

Diffluence (also spelled 'difluence'), or directional divergence, is one component of divergence. It is the directional spreading of

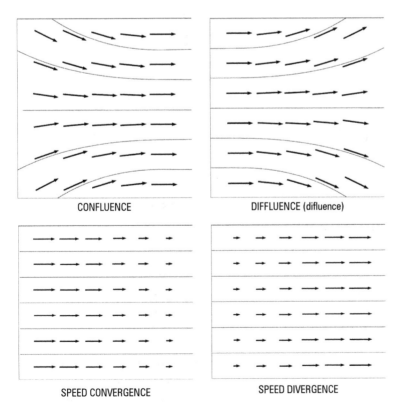

CONFLUENCE

DIFFLUENCE (difluence)

SPEED CONVERGENCE

SPEED DIVERGENCE

Figure 4-8. Divergence and convergence types. Shown here are vectors representing wind flow and streamlines, which are solid lines parallel to the wind.

wind flow. So if speed of the wind is constant everywhere but the flow shows diffluence, divergence is occurring. The corresponding term for directional convergence is confluence.

Speed divergence is another component of divergence. It is easiest to illustrate by picturing a highway with three lanes of traffic going 50 mph, and ahead of those cars is three lanes of traffic going 70 mph. In a sense, "mass" is being removed from between these vehicles as they spread apart. In the atmosphere, this is known as speed divergence.

When looking at charts, both diffluence/confluence and speed divergence/convergence must be looked at separately to determine if divergence is occurring. For example, on the front side of a jet streak it is common for diffluence to be cancelled out by speed convergence. The best way to tell whether divergence is occurring is through careful examination of both elements, or to use a computer program (such as Digital Atmosphere) that mathematically analyzes the wind field.

4.5. Vertical motion

Many introductory weather books tend to gloss over vertical motion. However, this type of motion is critical because it directly affects what the weather is going to be like. Vertical motion tends to be one of the key quantities that meteorologists are trying to solve for. The main problem, however, is that in synoptic scale motion, winds aloft are on the order of 50 meters per second, while the vertical motion that produces clouds and rain is on the order of 1 centimeter per second. This kind of quantity is not practical to measure directly, so forecasters rely on other indicators to analyze and predict vertical motion.

Dines Compensation Principle states that there must be at least one level of nondivergence in the troposphere. This level is called the level of non-divergence (LND). It is usually around 550 mb, but can be highly variable depending on atmospheric stability. If convergence occurs above the LND, divergence occurs below it, and vice versa.

4.5.1. DYNAMICALLY-FORCED ASCENT. If upper-level divergence occurs, the troposphere will attempt to compensate by initiating rising motion (to fill the "void"). Increased convergence in the lower troposphere will usually result. This process is sometimes referred to as the "chimney effect".

When air is forced to rise vertically from the surface, it rises into regions where pressure is lower (remember that pressure decreases with height). As a result, the volume of the air parcel

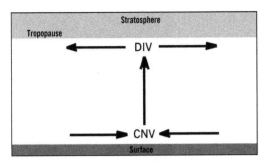

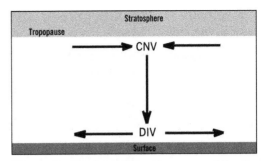

Figure 4-9a. The chimney effect. Divergence aloft and/or convergence at the surface will produce upward vertical motion (UVV's, or "upward vertical velocities").

Figure 4-9b. The damper effect. Convergence aloft and/or divergence at the surface will produce downward vertical motion (DVV's, or "downward vertical velocities").

expands, allowing its temperature to decrease. But eventually, the temperature will decrease to a point at which water vapor will be squeezed out of it. This produces visible cloud droplets, causing clouds. Continued rising motion produces more clouds, and the droplets can agglomerate into larger precipitation droplets, which are heavy enough to fall out of the cloud.

If the divergence aloft is stronger than the convergence in the lower levels, surface pressure falls will occur (mass is being removed from the column by the divergence).

If divergence aloft is strong enough, slight sinking motion may occur in the lower stratosphere, resulting in adiabatic warming. The result is warmer temperatures that are sometimes found above strong divergence areas (at 200 mb). This is called a "warm sink".

4.5.2. DYNAMICALLY-FORCED SUBSIDENCE. If upper-level convergence occurs, the troposphere will attempt to compensate by initiating sinking motion, or subsidence (to drain the "excess"). Increased divergence in the lower troposphere will usually result. This process is sometimes referred to as the "damper effect". The sinking motion produces adiabatic warming, and often clearing weather and fair skies.

If the convergence aloft is stronger than the divergence in the lower levels, surface pressure rises will occur (mass is being added to the column by the convergence).

Slight rising motion may also occur in the lower stratosphere, resulting in adiabatic cooling. The result is cooler temperatures that are sometimes found above strong convergence areas (at 200 mb). This is called a "cold dome".

4.5.3. VERTICAL MOTION FROM FRICTIONAL EFFECTS. In the boundary layer, frictional effects can produce vertical motion. For example, frictional convergence produces rising motion. This is a form of boundary layer convergence. An example is winds blowing across Lake Michigan and into Michigan itself; friction is less over the lake than on land, so the air slows over Michigan and convergence between the lake and the land wind results.

Likewise, frictional divergence is a form of boundary layer divergence and produces subsidence. An example of frictional divergence is wind coming out of a mountainous area and onto flat terrain. Friction decreases over the flat area and the wind speeds up. The area between the slow mountain wind and the fast plains wind is an area of divergence.

4.5.4. OROGRAPHIC FORCING. Any type of sloped land that deflects wind vertically produces vertical motion. This may be strong and localized, such as westerly winds impinging on the Coast Range and producing a locally intense band of ascent on the scale of minutes. It may also occur in the form of broad, weak motion, such as the westerly winds on the Great Plains which gradually sinks on a scale of hours or days. Broad, weak orographic forcing is generally referred to as "upslope" or "downslope" flow.

4.6. Jets

In meteorology, the word "jet" is a generalized name given to any narrow band of strong winds, ranging from the mighty jet stream which rings the globe to small scale inflow jets which occur on the back of a squall line. Because of this, it's always important to qualify the word to avoid any ambiguity. There are three primary types of jets in general meteorology work: the polar jet, the subtropical jet, and the low-level jet.

4.6.1. POLAR FRONT JET (PFJ). Out of all the different types of jets that exist, the polar front jet, or "polar jet", is by far the most prominent one in meteorology. It is caused by the temperature contrasts between polar and tropical air masses. These two air mass types have different densities and thicknesses, and as a result, they produce a height or pressure gradient aloft. The wind at these upper levels is forced to accelerate to maintain the balance of forces.

The PFJ flows west to east in both hemispheres, though some small segments may temporarily flow in the opposite direction. It circles the globe, forming a river of air known as the jet stream.

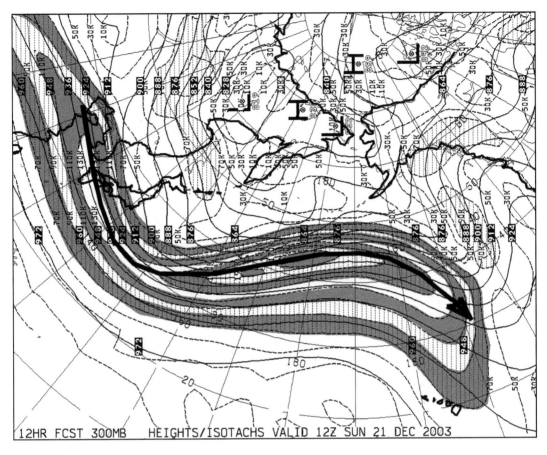

12HR FCST 300MB   HEIGHTS/ISOTACHS VALID 12Z SUN 21 DEC 2003

Because it is hemispheric in scope, a hot air balloon launched into the jet stream can be expected to flow from North America to Europe, then across Asia, then back over North America. The mean location of the jet stream migrates north during the summertime, when the extent of polar air masses recedes north, taking thermal contrasts with them. North American forecasters usually find the jet stream in Canada during the summertime, but in wintertime, the mean position shifts into the United States.

The PFJ represents a major source of energy for weather systems. For the jet stream to form in the first place, a boundary between polar and tropical air must exist. Once the jet stream strengthens, the jet can impart energy back to the surface systems, helping them develop further. The surface systems then can impart energy back to the upper levels. This chain reaction is known as self-development, and in extreme cases it can produce very intense extratropical cyclones.

4.6.2. SUBTROPICAL JET (STJ). Forecasters often find bands of strong winds in subtropical latitudes, such as along the border

Figure 4-10. Jet stream over the Pacific Ocean, with a large embedded jet streak south of the Aleutian Islands. *(21 Dec 2003 / 1200 UTC)*

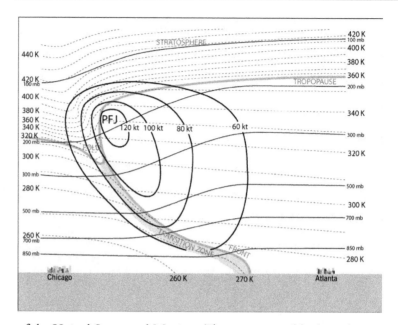

Figure 4-11. Conceptual cross section of a polar front jet. Isentropes are dashed thin lines, contours are solid medium-weight lines, and isotachs are thick solid lines. The key concept illustrated here is that the jet stream lies almost entirely above a frontal surface and is poleward of the surface front. Another concept shown here is, when moving from warm to cold air, the sudden upward slope in the isentropes once the front is encountered. If the wind is flowing in this direction, then it will ascend, resulting in widespread upward motion.

of the United States and Mexico. These jets resemble the polar front jet but have a different thermal structure and different characteristics. Known as subtropical jets (STJ) they tend to occur strictly at the upper troposphere and rarely extend downward into the middle troposphere. Because the tropopause is higher in the subtropics, they may be seen only on 200 or 100 mb charts. However they are important features and can easily produce winds of over 150 kt, resulting in well-ventilated, long-lived thunderstorms in regimes where no fronts or textbook short waves are present.

The STJ is caused by slight temperature contrasts along the boundary of the Hadley and Ferrel cells in subtropical latitudes. Where segments of the STJ bulge into higher latitudes, air parcels flow through the jet axis at a higher velocity. This is because angular momentum, which is a conserved property, is: $L = mvr$. Thus if the radius of the Earth decreases, which is true at these higher latitudes, velocity must increase. The air accelerates as a result.

The STJ is not always present. Where it occurs, its mean latitude averages about 28 deg. It is strongest in the winter, and tends to disappear during the summer. The STJ is a key player in transporting tropical air northward in the winter months, which helps enhance mid-latitude weather systems. Cirrus is often found on the warm side of the STJ axis, which tends to forms transverse bands and can help identify the jet's location.

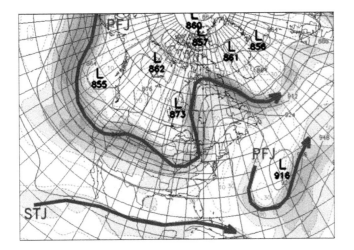

Figure 4-12a. Single stream pattern. A single polar front jet stream spans the North American continent. A weak subtropical jet exists over Mexico, but the pattern remains a single stream pattern. Out in the Atlantic it resembles more of a split. *(4 April 2010 / 0000 UTC)*

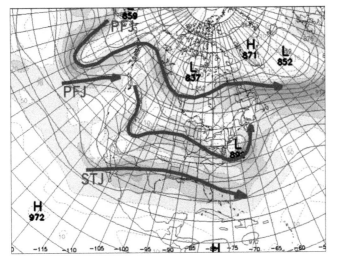

Figure 4-12b. A split flow pattern, which is more often the rule than the exception. This map shows two polar jets, one over Canada and another over the United States. There is also a prominent subtropical jet over the Gulf Coast states. *(24 March 2010 / 0000 UTC)*

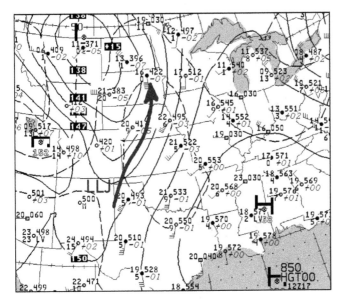

Figure 4-13. Low level jet. Low level jets are highly effective at transporting rich moisture into severe weather threat areas. This example is from the morning of an unprecedented tornado outbreak in Minnesota, with 48 twisters touching down. *(17 June 2010 / 1200 UTC)*

4.6.3. Low-level jet (LLJ). The low-level jet is a band of unusually strong poleward-moving winds (southerly in the northern hemisphere) in the lower troposphere. They are usually seen ahead of developing cyclones on the Great Plains. They are strengthened by the indirect transverse circulation ahead of a polar front jet, a process discussed in detail later in this chapter.

Low-level jets are strongest at 850 mb or 925 mb during the overnight hours. The typical location is from east Texas northward to Kansas or Missouri.

Low-level jets are important in advecting large volumes of warm, moist air northward from the Gulf of Mexico into the central United States. This process is a major factor in the release of severe thunderstorms later during the afternoon and evening hours.

As the LLJ may be co-located with isentropic lift where it crosses over cooler air masses, it is frequently associated with stratiform clouds, in particular stratus and stratocumulus. Cumuliform clouds may occur when solar heating sets in, and as these clouds build they may either spread into stratocumulus, dissipate, or build into thunderstorms, depending on the atmosphere's instability profile.

4.7. Jet streaks

A locally intense portion of a jet is referred to as a "jet streak" or a "jet max". Most jet streaks on the polar front jet have a scale of tens or hundreds of miles. They are found by analyzing isotachs, contours of wind speed and finding the telltale concentric patterns elongated along the jet axis. Jet streaks are important because the winds around it tend to be unbalanced, and this results in important vertical motion.

The jet streak's entrance region, or rear side, is upstream, usually to the west. Here air flowing into the jet streak is moving relatively slowly. However, the height gradient which actually causes the jet streak is tightening in this area. The height gradient force causes the air parcels to turn to the left (right in SH) and toward lower heights. At this point, the height gradient force exceeds contributions from the Coriolis effect. Since the air is moving slower than it would if it were in geostrophic balance, the air is considered to be "subgeostrophic". As the air moves toward the lower heights, it accelerates, gaining kinetic energy. This increase in velocity causes the Coriolis term to increase, and soon the air turns more toward its original direction, and is back in geostrophic balance.

When air exits the jet streak core, in other words, downstream of the jet streak, or at its front side, it is moving at a high velocity. The Coriolis effect is strong due to the contribution of velocity in the Coriolis term. The height gradient in the exit region begins weakening. But the air has momentum and continues moving rapidly, so the Coriolis effect is dominant. It turns more toward the right (left in SH) and in flowing away from lower pressures begins losing kinetic energy. This air is flowing faster than dictated by the height gradient, so it is said to be in "supergeostrophic flow". Eventually the velocity of the air decreases, the Coriolis force loses influence, and the balance of forces is restored. The parcel resumes its original direction of movement.

4.7.1. Divergent/convergent quads. The ageostropic (non-geostrophic flow) at the entrance and exit of the jet streak produces some interesting and very important effects. It produces divergent and convergent quadrants around the jet streak, which cause vertical motions. If we assume that the air surrounding the jet streak is all moving in a uniform direction and speed, it holds that the discontinuities in the jet streak where the air turns to the right and to the left creates "pockets" of convergence and divergence.

We will use the metaphor of a broad highway with 20 lanes. The outer lanes are moving at 50 to 60 mph, with the center lanes moving at 60 to 70 mph. In the midst of all this, an airplane is following along at 50 mph attempting to find a gap in which to land. The cars in the center lane react to the airplane, speeding up to 90 mph as they approach it to avoid being landed on, and after getting past it, they slow back down. All cars also have a defect where the steering wheel pulls to the left when they accelerate.

Using this example, it's easy to picture the traffic collisions that will occur to the left rear and front right of the airplane as the

Figure 4-14. Jet quadrants and their meanings in the Northern Hemisphere. The left front and right rear quadrants have the greatest association with clouds and precipitation.

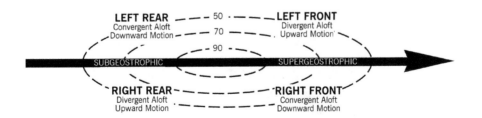

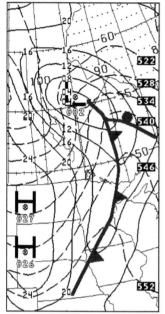

Figure 4-15. Thickness patterns, shown here as dashed lines, look surprisingly like isotherms. This isn't a coincidence: thickness is functionally like an "average isotherm" throughout the layer, in this case 1000-500 mb (surface to 18,000 ft). Therefore they shouldn't be considered intimidating. The sea-level pressure field is superimposed to help determine areas of cold and warm advection. Note how these two fields form "boxes" where advection is taking place. The smaller the box, the stronger the advection. Strong cold air advection is taking place behind the cold front, while strong warm advection is occurring north of the warm front. This example is for the morning of May 4, 2002.

accelerating traffic collides with other cars which are not reacting as much to the airplane. There will also be voids to the right rear and left front of the airplane where no cars are found. This is an example of mass convergence and divergence, respectively, taking place in the upper troposphere.

As a result, the mass divergence in the right* rear quadrant and left* front quadrant will create compensating upward motion, destabilizing the atmosphere and perhaps producing clouds and rain. The mass convergence in the left* rear and right* front quadrants will be associated with compensating subsidence, favoring fair skies.

4.7.1. TRANSVERSE CIRCULATIONS. These areas of upward and downward link tend to link across the jet axis (left and right, or −*n* and +*n*) to form transverse circulations. For example, if we analyze the flow around the jet in detail, we would notice a component of motion in which air subsides on the right front quadrant of a jet, sinks to the low levels, flows along the surface to the area of rising air under the left front quadrant, rises, then flows aloft back to the convergent right front side. This is one example of a transverse circulation, and it is considered *thermally indirect*, since it flows opposite of the dense, cold sector's tendency to sink and the warm air's tendency to rise. Since it works against the natural thermal circulation, kinetic energy is being converted into potential energy.

In the entrance (rear, or upwind side) of a jet streak, the air has a component of motion where it tends to sink on the left side, flow along the surface to the right side, rise, then flow aloft to the left side. This is a *thermally direct transverse circulation* since it works *with* the air's natural tendency to sink in the cold dense side and rise on the warm buoyant side. Potential energy is being converted into kinetic energy.

4.8. Thermal advection

Just as warming or cooling from diabatic or adiabatic sources can cause changes in the atmosphere aloft, the mere replacement of warmer or cooler air by wind transport can also produce changes. This type of transport is called *advection*. Cold and warm air advection can be readily estimated using isotherms, winds, and thickness charts.

Different signs of advection (i.e. warm and cold) in two different layers in the vertical produces an effect called *differential advection* which stabilizes or destabilizes the atmosphere. For example, if there is low-level warming and upper-level cooling,

---

* Asterisks from here on in the text indicate a quantity that this is true in the Northern Hemisphere and should be reversed in the Southern Hemisphere.

the temperature contrasts with height are steepened, the lapse rate increases, and the atmosphere destabilizes.

It should be cautioned that subtle advection, especially that which occurs aloft, is not easily diagnosed using constant pressure or constant height charts, because the flow usually has a vertical component, and adiabatic and diabatic temperature changes may overpower changes from advection. If advection is strong and focused near the surface, it is usually safe to estimate it using isotherms, but otherwise, it's often safer to use isentropic charts and cross sections, described later, to assess advection.

4.9. Thickness

Not surprisingly, a toy balloon left outside on a cold night will shrink. This is because the volume a parcel of air occupies is proportional to its temperature, and to a lesser extent, its moisture. All of the air parcels that make up a column of air in the troposphere respond in this way to changing temperature and moisture, just like a room filled to the ceiling with toy balloons. The highest balloons in the stack undergo the greatest vertical movements because there are more balloons underneath responding to the temperature changes. In a sense, thickness measures the distance between layers of these balloons.

In meteorology, thickness is an expression of the vertical distance between two constant pressure surfaces. When mapped

Figure 4-16. Thickness charts are valuable for locating frontal systems, especially in the mountains. Since thickness is effectively a layer average of temperature, the patterns may not correspond precisely to the location of surface fronts, but air mass contrasts are still easy to see, nevertheless. Here we see the December 1998 cold air outbreak, which brought the seventh coldest air outbreak on record to the Great Basin region. A potent baroclinic cyclone is developing in Nevada with a deep cold air mass in the Pacific Northwest. Just as fronts are found on the warm side of isotherm ribbons, they are also found on the warm side of isothick lines as shown here. Where isothicks and isobars form a webwork of boxes, thermal advection is strongest where the boxes have the smallest area. *(19 Dec 1998 / 1200 UTC)*

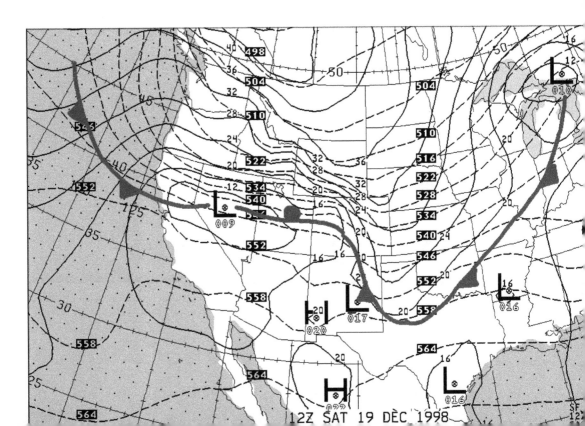

out across a region, it shows the "mean temperature" of a layer. Lines of equal thickness are isopachs, but due to the obscurity of that term they are sometimes simply called "isothicks".

4.9.1. LAYER SELECTION. The most common top and bottom surfaces for a thickness layer are 1000 and 500 mb, which yields a quantity known as the 1000-500 mb thickness. Since thickness is a distance quantity, the units preferred are meters, preferably decameters (dam). A typical 1000-500 mb thickness is 548 dam (5480 m, or about 3 miles. Other layers can be selected, also. Using too thick of a layer may average out unwanted data. For example, if a forecaster is looking for a shallow front, a 1000-500 mb thickness chart will be biased by the extensive depth of warm air aloft from 900 to 500 mb overlying the front. The forecaster should select the right layer heights and the right layer depth to find the desired features.

When using a 1000-500 mb thickness chart, contour intervals of 6 dam (60 m) from a base of 558 dam should be used. The estimated position of the polar front jet based on thickness gradient is drawn as a thick black arrow. The estimated position of surface fronts based on thickness gradients are drawn using standard conventions.

4.9.2. RELATIONSHIP TO DENSITY. Since thickness is an average of the thermal properties within the column of air, it can be thought of as an expression of the "mean temperature" of the layer. More properly, this is an expression of "mean density", since moisture affects the thickness also, but temperature is the main contributing value. Moist air biases the thickness value downward, while dry air biases it upward. So a moist, slightly cool air mass will exhibit about the same thickness as a dry, slightly warm air mass. This is one reason why a dryline in Texas rarely shows a thickness gradient, even though temperatures across it may differ.

Forecasters often combine the temperature and moisture into an expression of density known as virtual temperature. In the

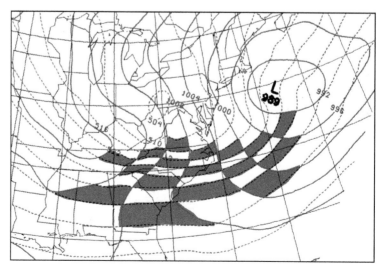

Figure 4-17. The advection box method is similar to the techniques for advecting vorticity with the box method, analysts can form boxes of thermal advection using the isobar field and the thickness field. The smaller the box, the stronger the advection. Shown here is a region of cold air advection in the wake of a cold front on the East Coast. *(9 January 2011 / 0000 UTC)*

example above, a moist, slightly cool air mass and a dry, slightly warm air mass will have the same virtual temperature and the same thickness.

4.9.3. THICKNESS AND THERMAL ADVECTION. Charts showing thickness can be used to determine thermal advection of that layer. The presence of warm air advection (WAA) is indicated where geostrophic winds (winds along isobars or isoheights) for that layer blow higher thickness values into a region occupied by lower thickness values. The presence of cold air advection (CAA) is indicated where geostrophic winds carry lower thickness values towards a region occupied by higher thickness values. Neutral advection occurs if the geostrophic wind is weak (weak pressure gradient), a weak thickness gradient exists, or the geostrophic wind blows parallel to the thickness contours.

This leads us to the "box technique". On a chart where sea level pressure isopleths are meshed with thickness isopleths, advection can be determined by visually forming boxes created by the intersection of the two isopleth types. Areas with lots of small boxes always correspond to both a strong pressure gradient and a strong thermal gradient, and this implies very strong thermal advection. Areas where the boxes are large or do not close off are areas of weak thermal advection. All the forecaster has to do is determine the type of thermal advection that is taking place and mark or color it appropriately. It is best to select a mean layer wind that approximates the thickness layer. It should be pointed out, though, that a box method based on 850 mb heights or surface pressure is just an approximation as it only considers advection of the surface portion of the thermal layer.

4.10. Frontogenesis and frontolysis

A front is simply a zone of thermal contrast, so it follows that frontogenesis is a strengthening of this contrast and frontolysis is a weakening. It is the orientation of the thermal field with respect to the wind field which largely determines whether thermal gradients will amplify. The most familiar cause of frontogenesis [frontolysis] is the convergence [divergence] of air. This moves thermal gradients closer together [further apart]. This convergence [divergence] can result from either confluence [diffluence], speed convergence [speed divergence], or a combination of the two.

If a thermal field is tilted into the vertical, a thermal gradient strengthens. In a thermally indirect circulation, such as in the exit region of a jet max, cold air is rising in the cold sector and

Misuse of diffluence

Diffluence is simply the spread of streamlines downstream . . . The diffluence is given only by part of the first term [for divergence], therefore, the diffluence of the flow cannot be equivalent to the divergence. It is a common mistake, however, to equate diffluence with divergence (and then go on to infer vertical motion, another mistake).

CHUCK DOSWELL,
"Pet Peeves of Chuck Doswell,"
2000

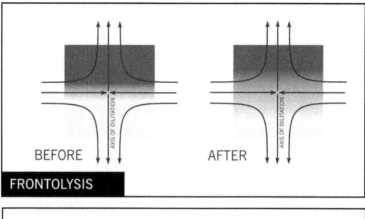

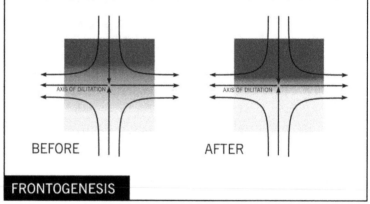

Figure 4-18. The effect of a deformation zone on the thermal gradient. What occurs is predicated on two things: the fact that in a deformation zone divergence occurs along one axis and convergence along the other axis, and that only the axis that is perpendicular to the thermal gradient has an effect on it (i.e. pulling a mass of cheese apart makes it wider but not necessarily thinner). If the axis of dilitation is perpendicular to the thermal gradient (top), then the divergence axis is dominant and frontolysis occurs. If the axis of dilitation is parallel to the thermal gradient (bottom) then the convergent axis is dominant and frontogenesis is the result.

warm air is sinking in the warm sector. This produces adiabatic cooling in the cool sector and adiabatic warming in the warm sector. It tilts the isentropic surfaces into the vertical and results in frontogenesis. In a thermally direct circulation, the opposite effect is taking place, and this "untilts" the gradient and leads to frontolysis.

If a deformation zone exists, the angle between the axis of dilatation and the isotherms determines what is taking place. If the isotherms are parallel [perpendicular] to the axis of dilatation, then the inflow is convergent [neutral] and outflow is neutral [divergent], resulting in frontogenesis [frontolysis].

Finally, diabatic heating can have an influence on a frontal zone. If the cloud cover fields cause solar heating to concentrate in the warm [cold] sector, this intensifies [weakens] the thermal gradient and causes frontogenesis [frontolysis].

One pattern forecasters can look for which indicates frontogenesis is strong wind shear and a straight-line hodograph or unidirectional flow. The former suggests a baroclinic zone and the latter suggests the thermal zone has vertical consistency with respect to direction.

4.10.1. FRONTOGENETICAL FORCING. A baroclinic zone, which is essentially a density gradient, produces a pressure gradient aloft, which causes a jet streak to form. This produces ageostrophic flow which is out of balance. The four-quadrant jet concept described earlier is one example of circulations that exist in an ageostrophic situation.

In a typical mid-latitude cyclone, the strongest frontogenetic forcing tends to occur just poleward and west of the surface cyclone, which is also part of the deformation zone cloud pattern and in the vicinity of the left-front quadrant of the jet.

The precipitation and weather that develops is largely modulated by stability. In a stable atmosphere, the precipitation tends to be widespread and weak, while in an unstable atmosphere, intense bands are favored which tend to align along isentropes. Stability considerations include not only gravitational stability but especially symmetric instability.

4.10.2. RESPONSE. When the atmospheric forces are out of balance, the atmosphere develops a *response* to restore the balance. With frontogenetical forcing, the response is the development of a direct thermal circulation, with sinking motion of dense air in the cold sector and rising motion of buoyant air in the warm sector. Since the frontal surface and the isentropes slope upward towards the cold air, the direct thermal circulation decreases the slope of the frontal surface, produces adiabatic warming in the cold sector, adiabatic cooling in the warm sector, weakens the thermal gradient, and leads to frontolysis.

4.11. Vorticity

From the 1960s through the 1980s, vorticity advection was used extensively to forecast vertical motion, because modelling of vertical motion was rather limited and vorticity fields were readily available on the limited assortment of charts that were available. It was found that cyclogenesis was found downstream from areas of cyclonic vorticity, with the opposite true for anticyclonic vorticity. Areas where advection is bringing in higher values of vorticity are known as positive vorticity advection (PVA) areas, with the reverse true for negative vorticity advection (NVA). In the Northern Hemisphere, PVA is equivalent to CVA (cyclonic vorticity advection) and NVA to AVA (anticyclonic vorticity advection).

The basic forecast process uses 500 mb absolute vorticity with height contours overlaid, representing streamlines. The

Garlic roast potatoes
5-6 medium red potatoes
1/2 c. olive oil
1/2 c. melted butter
2 tbsp garlic powder
1/4 tsp red pepper powder
1/4 tsp salt
1/4 tsp coarse black pepper
2 dashes cinnamon powder
dash paprika
Preheat oven to 400°F and grease baking sheet lightly. Scrub potatoes and cut, unpeeled, into 1-2 inch chunks. Toss everything in deep bowl until all potatoes are coated. Distribute coated potatoes on pan (save sauce) and bake. Every 20 minutes remove from oven, drain excess oil, brush liberally with more sauce, and return to oven. Total baking time is 45 to 70 minutes or until potatoes are slightly charred.

intersection of vorticity contours and height contours forms distinct boxes, and where the boxes are smallest, vorticity advection is the most intense. This is similar to the technique used for interpreting thermal advection from thickness charts. If the charts show CVA, upward motion is suggested, with possible surface cyclogenesis and deteriorating weather, while AVA suggests downward motion, surface anticyclogenesis, and improving weather.

For CVA/AVA to work as indicators of vertical motion, the winds have to be nearly geostrophic. Unfortunately much of the significant weather that occurs is produced by winds that are out of geostrophic balance. Also CVA cannot be offset by cold air advection or AVA by warm air advection. Furthermore, it has to be assumed that the vorticity patterns are not moving faster than the upper-level winds, and that under the given circumstances 500 mb is in fact fairly close to the level of non-divergence. Additionally, the omega equation for vertical motion shows that CVA or AVA must increase with height for vertical motion to result. This adds further complications for the forecaster. The implications of this are that CVA and AVA should be used strictly as one of many possible indicators for which type of vertical motion is taking place.

It has been shown that CVA/AVA and thermal advection tend to cancel each other out. Another approach to the problem has been to integrate both parts of the omega equation through evaluation of vorticity patterns by the thermal wind. This can

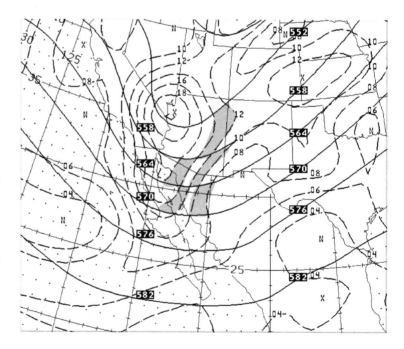

Figure 4-19. Vorticity advection strength can easily be estimated by superimposing the vorticity field on the height field and looking for "boxes" formed by the intersection of lines. The smaller this box is, the stronger the vorticity advection, and presumably the stronger the vertical motion. A strong area of cyclonic vorticity advection (PVA) is indicated by the shaded boxes shown here. This area is likely to contain clouds and possibly precipitation. Further west, another area of boxes exists in an area of anticyclonic vorticity advection behind this short wave trough, but are not drawn on this chart. (30 December 2002 / 000 UTC)

be done by using the thickness contours rather than the 500 mb height contours to determine advection of vorticity. The advection of vorticity patterns through this technique is called PIVA and NIVA (positive and negative isothermal vorticity advection). Some informal studies have shown that using PIVA/NIVA has clear advantages, but little has been published on it.

4.11.1. VORTICITY AND JET STREAKS. The convergent and divergent areas around jet streaks can be illustrated using vorticity principles. As we describe jet streaks, we will be using the left/right and front/rear coordinate systems, which refers to directions as if we are sitting on the jet stream or on a jet streak, straddling the axis and looking downstream. Therefore "left" is generally north* of the jet axis, and "rear" generally refers to a westward direction (since our back faces west).

For a jet streak in straight-line flow (Figure 4-21) all vorticity is produced by shear. The left side experiences cyclonic* shear and a vorticity maximum is produced. The right side experiences anticyclonic* shear, and a vorticity minimum is produced. The right rear quadrant experiences PVA since it is behind the vorticity min, and thus upper divergence* is indicated. The left

Figure 4-20. A 500 mb vorticity analysis showing the two primary types of vorticity lobes. Channel lobes, as seen here off the Pacific Coast, are elongated parallel with the wind, and the vorticity centers are associated primarily with shear. A jet streak is found between the channel lobes (shaded here along the jet axis). On the other hand, advection lobes, as seen here over Arizona, are associated primarily with curvature, and are generally a manifestation of troughs and short waves. Note that the jet stream crosses contours somewhat near the jet streak exit; this is an example of supergeostrophic flow. This date was on the morning of May 3, 1999, when Oklahoma's largest tornado outbreak occurred. *(3 May 1999 / 1200 UTC)*

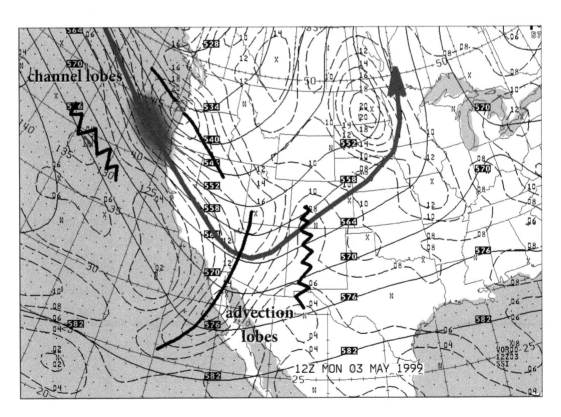

Figure 4-21. Vertical motion and vorticity associated with jet streaks. The left panel shows vertical motion, and the right panel shows vorticity contours and patterns. Isotachs are depicted by the medium-weight gray lines, with highest wind speeds at the center of the concentric patterns.

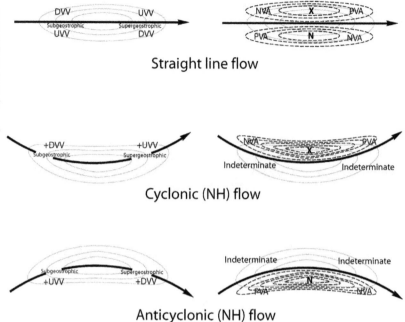

rear quadrant experiences NVA since it is behind the vorticity max, and thus upper convergence* is indicated. The left front quadrant experiences PVA since it is ahead of the vorticity max, and thus upper divergence* is indicated. The right front quadrant experiences NVA since it is ahead of the vorticity min, and thus upper convergence* is indicated.

For a jet streak in cyclonic* flow (again, Figure 4-21) vorticity patterns are produced by both shear and curvature. The left side experiences both cyclonic* shear and cyclonic* curvature, so a strong vorticity maximum is produced. The right side experiences anticyclonic* shear and cyclonic* curvature, which negate each other. The right rear quadrant is indeterminate since there is no max or min ahead of it. The left rear quadrant experiences strong NVA since it is behind the strong vorticity max, and thus strong upper convergence* is indicated. The left front quadrant experiences strong PVA since it is ahead of the strong vorticity max, and thus strong upper divergence* is indicated. The left right quadrant is indeterminate since there is no max or min behind it.

For a jet streak in anticyclonic flow (again, Figure 4-21), vorticity is also produced by both shear and curvature. The left side experiences cyclonic* shear and anticyclonic* curvature, which negate each other. However the right side experiences anticyclonic* shear and anticyclonic* shear, which add up to strong anticyclonic* vorticity. The right rear quadrant experiences strong PVA since it is behind a strong vorticity min, and thus strong upper divergence* is indicated. The left rear quadrant is indeterminate since there is no max or min ahead of it. The left front quadrant is indeterminate since there is no max or min behind it. The right front quadrant experiences strong NVA since it is ahead of a strong vorticity min, and thus strong upper convergence* is indicated.

4.11.2. VORTICITY PATTERNS. Vorticity contours may elongate into either shear lobes or advection lobes, depending on how they are aligned with the wind flow.

Shear lobes are elongated parallel to the wind, and typically occur left and right of elongated jet streaks, sometimes forming a couplet on either side of the jet streak itself. Segments of the jet stream which are associated with shear lobes are sometimes called channel jets. Since the vorticity contours of a shear lobe tend to align with the height contours, weak vorticity advection is indicated.

Advection lobes are elongated areas of vorticity which are aligned largely perpendicular to the wind flow. These are usually

Figure 4-22. Vorticity patterns in weak flow are not associated with strong vertical motion, because advection is weak. The vorticity values by themselves mean little without the presence of strong flow. A chart like this is common in the summer months in the United States. *(1 August 1998 / 0000 UTC)*

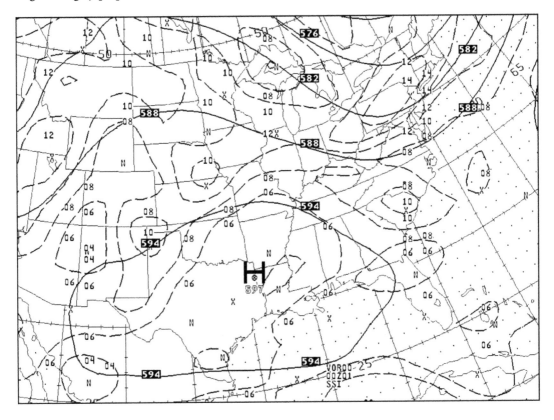

what are referred to when a forecaster speaks of a "vorticity lobe". Advection lobes occur within a jet stream where curvature is somewhat localized and sharp, such as within a short wave. Therefore curvature and not shear predominates in the vorticity equations. Advection lobes generally indicate the possibility of a short wave trough or ridge at 500 or 700 mb and should be checked out further.

Jet streaks which are associated with advection lobes are called advection jets, though since the main contribution is from curvature, there may be no discernible jet streak. Vorticity advection occurs along the jet axis with a strong vorticity advection gradient, so dynamics tend to be strong.

4.11.3. ISOTHERMAL VORTICITY ADVECTION. Vorticity contours are traditionally overlaid on depictions of pressure or height. However if we consider advection by the thermal wind field, this more closely relates the parcel to vertical motion. Fortunately, the thickness chart lends itself well to depicting the thermal wind. However, the vorticity chart should be in the middle of the layer; thus the 500 mb vorticity chart should be in a 700-300 mb thermal layer, and the 1000-500 mb thickness should use 700 or 850 mb vorticity.

4.12. Q vectors

A Q vector is a derived measure of vertical motion. It highlights areas where the thermal wind balance is being disrupted, which will cause compensating vertical motions. The Q vector approach combines the two key contributions to vertical motion: temperature advection and convergence/divergence due to changes in vorticity advection with height. Q vectors are interpreted as follows: Q vector convergence [divergence] indicates rising [sinking] motion at that level. If Q vectors point toward warm [cold] air, frontogenesis [frontolysis] is indicated.

The technique is good because vorticity advection, the indicator that has long been used by forecasters for decades, is often cancelled out by thermal advection. Likewise, vorticity advection can negate the anticipated effects of thermal advection. Q vectors provide a quantitative way of measuring both elements. They cannot be easily computed by hand, and the output is quite difficult to find (especially on the Web), but certain software programs and weather analysis systems can easily calculate the values.

On vertical motion diagnosis
"If one is computing the forcing terms in the omega [vertical motion] equation numerically, it is best to compute the divergence of the Q vectors. One advantage of the Q vector approach is that the vectors themselves have physical significance, providing a visualization of the low-level branch of the ageostrophic circulation which supports the vertical motion. Another significant advantage is that the computation of the divergence of the Q vectors is subject to less numerical error.

DALE DURRAN & LEONARD SNELLMAN, 1987

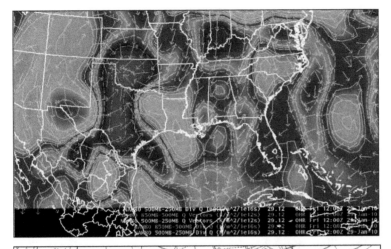

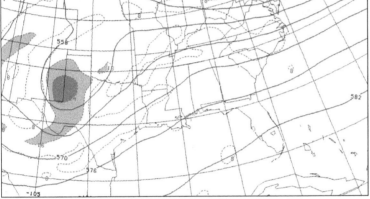

Figure 4-23. Q vectors and Q vector convergence as shown on AWIPS (top) and on the NCEP 500 mb heights and vorticity panel (bottom). The traditional heights and vorticity method indicates upward motion from the Abilene area northward to southwest Oklahoma. The Q vector fields indicate a similar profile for Q vector convergence. *(29 Jan 2010 / 1200 UTC)*

4.13. Isentropic analysis

Isentropic analysis refers to the diagnosis of vertical motion using isentropic surfaces (three-dimensional surfaces of equal potential temperature). Interestingly during the 1930s and 1940s, isentropic analysis was extremely common, yet fell into steep decline over the next several decades. It was only in the 1990s when interest in the technique was revived. Isentropic charts are most valuable during the winter season when diabatic effects (strong heating, etc) are less likely to corrupt isentropic analysis and interpretation.

Isentropic charts give forecasters an edge because parcels are not bound to constant pressure surfaces or constant height surfaces. They naturally move along potential temperature (theta, or θ) surfaces unless some process is occurring that adds or removes heat. The parcel is bound to the theta surface because if it is forced downward [upward], it will warm [cool] adiabatically and become warmer [cooler] than the air around it, making it relatively buoyant [negatively buoyant] compared to the environmental air; this causes it to rise [sink] until it returns to its original theta surface.

Again, the assumption is made that there are no diabatic processes affecting the parcel. A parcel that is saturated and is displaced upward will gain latent heat, causing it to cease moving along theta surfaces. Likewise a parcel that gains or loses heat due to some outside process such as conduction changes its temperature and in doing so changes the theta surface it is bound to.

Figure 4-24. Conceptual model of isentropic surfaces between a cold air mass and a warm air mass. The presence of a sharp frontal boundary is neglected to help simplify this example. Note how at the ground, the temperature decreases from about 289 K to 278 K. This causes an upward "bulge" where the cold air mass exists. Therefore if air is blowing from the right to the left side of the page, then parcels are forced to ascend to remain on their adiabatic surface. Ascent or descent is determined relative to the pressure contours (gray) rather than to actual height or elevation. The pressures surfaces sag over the cold air mass due to the lower thicknesses present there.

4.13.1. ISENTROPIC SURFACES. Forecasters must remember that potential temperature (represented by θ or adiabats on the skew T diagram) always *increases* with height. For example, a typical theta value will be about 290 deg K near the ground, increasing to 330 deg K in the upper troposphere. Potential temperature decreasing with height implies an absolutely unstable layer, i.e. cold over warm, which will immediately overturn and mix out.

Isentropic surfaces are created by connecting points of equal potential temperature. These surfaces form layers, like a thick stack of blankets one atop the other, with the highest layer corresponding to the highest potential temperature. These surfaces form domes over cold air masses, because the cold air masses harbor cold potential temperatures compared to the warm air.

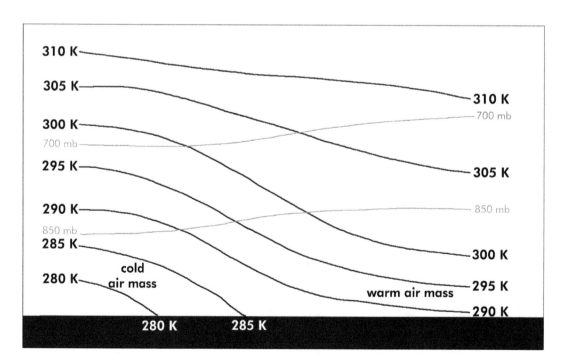

Isentropic surfaces are labelled in units of Kelvin, so a parcel at 1000 mb with a temperature of 32 deg C, which is 305 K, will be found on the 305 K surface.

Looking at the isentropic surfaces, the spacing of surfaces in the vertical is related to how stable or unstable the air mass is. Remember that potential temperature always rises with height. If there are many isentropic surfaces packed close together, this indicates strong warming with increasing height, thus a warm-over-cold or stable situation. Large spacing between surfaces indicates relatively unstable air.

4.13.2. MOTION. Air tends to cling to the potential temperature surface it is located at. This is because potential temperature is conserved; it can never be changed unless heat is added to or removed from the parcel. Therefore air tends to follow isentropic surfaces. Again, going back to the analogy of being in bed, an air parcel would be trapped between the sheets. Your body, representing a cold air mass, bends the surfaces upward. If the air is blowing from one side of the bed to the other, the parcel will be forced to rise as it moves toward the cold air mass. This is called "isentropic lift" or "isentropic upglide", and describes exactly what happens as the low-level flow moves poleward across a warm front. Likewise, when the air parcel leaves a cold air mass that is stationary, it descends to lower surfaces. As with any air that is rising, parcels undergoing isentropic upglide may saturate and produce clouds and precipitation.

Forecasters must be attentive to whether diabatic processes are taking place within a layer, because this will cause parcels to cease moving along isentropes. For example, if a parcel is saturating, it is gaining latent heat and will become buoyant, causing it to rise. In doing so, it is actually conserving equivalent potential temperature, not potential temperature.

It should be noted that if a layer is saturated , diabatic parcels may no longer conserve potential temperature.

4.13.3. ISENTROPIC CHARTS. The isentropic chart is a horizontal plot of height, wind, temperature, humidity, and other variables on a specific isentropic surface. Height or pressure contours are almost always plotted. By relating the wind plots to the height or pressure contours, we can readily see where air is ascending or descending. Plots of relative humidity also tell how close the air is to saturating and forming cloud material.

To use constant potential temperature maps, you have to figure out which isentropic surface you want to look at. If too low of an isentrope is selected, you'll be too close to the ground

Isentropic analysis websites
Unfortunately, as of 2011 isentropic analysis remains largely neglected by most weather sites on the web, though there are a few notable exceptions such as the College of DuPage and OU Oklahoma Weather Lab weather pages.

or underneath it, and much of the chart will be unusable. If it's too high, you'll be too high up and you may be missing what is happening in the lower and middle layers of the atmosphere. A rule of thumb is to locate the warmest sea-level temperature in your area of interest, convert it mentally to a Kelvin temperature, and select that as the isentrope. You can also look at cross-sections through your area of interest to select a good isentrope to look at. Usually in winter a good surface is 290-295 deg K, and in summer a good surface is 310-315 deg K.

4.13.4. CROSS SECTIONS. Instead of being bound to a semi-horizontal map, forecasters may choose to create a vertical cross section to view the isentropes. These are typically not available on the Internet and require a special software package such as GEMPAK or RAOB, or must be hand plotted.

The cross section should always be done along the wind streamlines to best represent the path a parcel will follow. In other words, if the winds are uniformly out of the south, then the forecaster will construct a north-south cross section to show the isentropic path it will follow. Since wind direction tends to change in direction with height, it is often difficult to properly orient a cross section. It should be aligned with winds mainly at

Figure 4-25. Isentropic analysis for the 295 K level, as obtained from the College of DuPage website. Contours are hand-drawn as solid isopleths. *(24 March 2011 / 0000 UTC)*

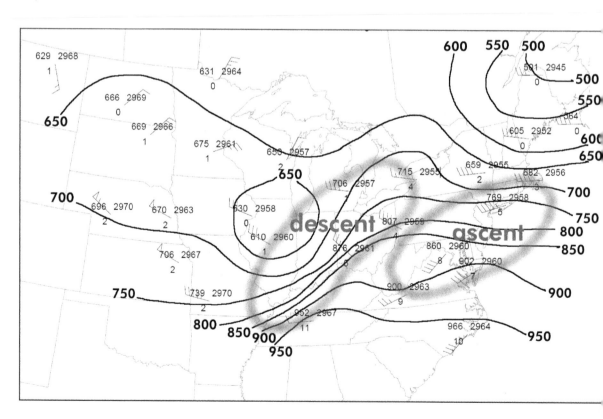

the level of most interest to the forecaster, such as the level with highest humidity.

Once a cross section path is established, the forecaster will plot isentropes and pressure contours. If the software permits, the wind component along the axis of the cross section can be plotted in order to show parcel motion. Other isopleths that are helpful are relative humidity, in order to determine how close the parcels are to saturation.

Chapter Four
REVIEW QUESTIONS

1. At which level are long waves most likely to be found?

2. What techniques can be found for locating a short wave trough?

3. What causes a jet stream to exist?

4. What is a thermally direct circulation, which portion of a jet streak is it found within, and is it frontogenic or frontolytic?

5. What type of weather occurs in the left front quadrant (northern hemisphere) of a jet streak, and why?

6. If the only chart you have is a heights and vorticity chart, what would be a good technique for locating a jet streak?

7. Why does potential temperature increase with height?

8. Are isentropic surfaces high or low over the tropics, compared to temperate latitudes?

9. If the winds are blowing south to north along a warm front surface, is this isentropic descent or ascent, and why?

10. Name a situation where isentropic analysis would be inappropriate.

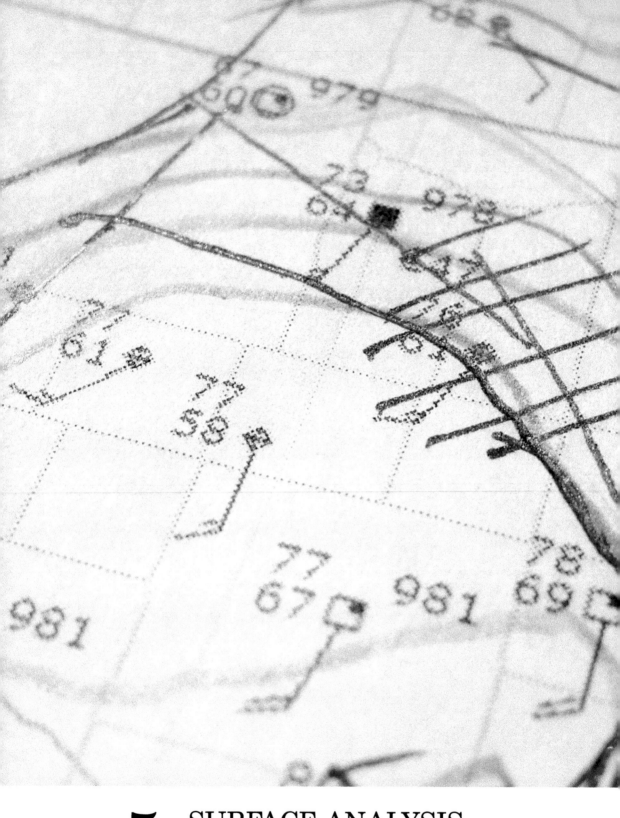

5 SURFACE ANALYSIS

S urface analysis focuses on weather conditions, air masses, and processes taking place close to the Earth's surface. While surface charts are often thought of as being the backbone of surface analysis, it also takes into consideration products like radar, satellite, and even looking out the window. A good surface analysis also considers elements from upper level data and soundings.

5.1. The surface chart

Although glancing at a weather map containing nothing but data plots might be sufficient to get an idea of what's happening, it's sometimes necessary to draw isopleths (lines) to establish the shape of the most important fields. It's common for computer software or Internet-based graphics to provide these isopleths, however they do tend to smooth out important patterns or features, requiring re-analysis by hand.

5.1.1. STATION PLOTS. The individual weather observations from each station are placed on the map as "packets" of information. These are known as station plots. Learning to read station plots, particularly those of surface reports, is an essential forecasting skill. A section is presented in the appendix that explains how to read station plots.

5.1.2. ISOPLETHS. The forecaster should sketch isopleths when time permits. This process does not simply accomplish an arcane task of drawing lines; rather it helps etch into the forecaster's memory all of the details of the atmosphere, and forces the forecaster to examine and ponder areas of unusual or unexpected readings. Surprisingly, as forecasters become experienced, many rely increasingly on hand analysis because of its proven value in finding key features.

Isoplething is done by first selecting a type of data that is most relevant to the forecast. Not all types of data are necessary, however one or two choices are recommended. In the mid-latitudes, isobars will be the most common choice, while in the tropics, streamlines are the norm. Some suggestions for contour base values and intervals are provided in the appendix, however the forecaster is free to choose an interval that yields the best results and helps important features stand out better.

Most importantly, the forecaster must resist the temptation to smooth the lines, that is, fudging isopleths where they don't accurately fit the data. The purpose of isoplething is to allow anomalies to stand out fully and be considered. Data values which are clearly erroneous may be circled by the forecaster and ignored, however all other data that is presumed to be accurate must also be isoplethed accurately.

5.1.3. ISOBARS. Lines connecting equal pressure are isobars. The centers of highest and lowest pressure are of greatest interest as they indicate the center of important weather systems. Elongations, troughing, or ridging with time into an area are also strong signals to the forecaster. Fronts always lie within pressure troughs, so the forecaster should expect to kink the isobars when drawing across a frontal boundary. Isobars constructed from altimeter setting values are sensitive to terrain elevation and don't work reliably in mountainous areas, but they are slightly more common in the U.S. than sea level pressure. Since 12-hour temperature corrections do not bias altimeter setting values, such isobars are much more responsive to mesoscale changes than sea-level pressure isobars.

5.1.4. ISOTHERMS. Lines representing temperature are called isotherms. Boundaries between warm and cold air, where isotherms pack tightly, can reveal the location of subtle frontal boundaries. However false gradients may appear due to elevation differences, station siting issues, or even by cloud cover. Wind flow perpendicular to an axis of isotherm packing suggests thermal

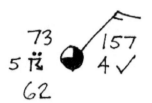

Figure 5-1. Surface plot model. This example shows a temperature of 73, a dewpoint of 62, with a visibility of 5 miles in a thunderstorm. The sea level pressure is 1015.7 mb, and the pressure tendency is 0.4 mb falling then rising. The sky cover (shading of circle) is 6/8ths, and the wind direction is northeast at 15 mph.

Figure 5-2. Surface analysis with isobars and other markings. This was used to prepare a severe weather forecast.

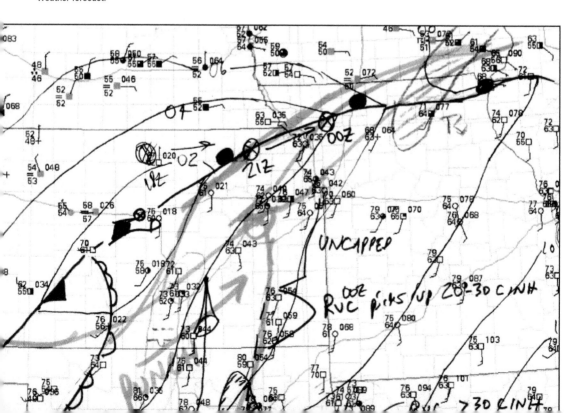

advection, and indicates that isentropic ascent or descent, and thus lift or subsidence, is probably occurring.

5.1.5. ISODROSOTHERMS. By isoplething the dewpoint temperatures the forecaster sees the surface moisture field. Before evaluating such fields it's important to realize that the chart does not provide information about the *depth* of moisture, so a 70-degree dewpoint in one part of the state might not have the same forecast impact as a 70-degree dewpoint in another part of the state. So moisture fields should be supplemented with soundings or charts at other levels above the ground. The axis of highest moisture is often, but not always, co-located with the deepest moisture. Forecasters, if they hand-analyze isodrosotherms, will usually just draw them in specific areas of interests.

5.1.6. ISOTACHS. Isotachs, which show values of wind speed, are an optional field that can be added. They can be used in high-wind events driven by strong pressure gradients, and the extent and trend of these isotachs can yield valuable forecast information. At the surface, isotachs are drawn every 5 or 10 kt in purple. They're generally only drawn in specific areas of interest.

5.1.7. OTHER MARKINGS. The forecaster is by no means limited to a specific set of markings. Anything that aids the diagnosis of the atmosphere should be added, including sketches from radar data, satellite, model forecasts, or elsewhere.

5.2. Air masses

As the surface analysis progresses, the forecaster begins thinking in terms of what kind of air masses cover various regions on the chart. Unfortunately, air mass analysis has been marginalized since the 1970s due to the heavy emphasis on numerical modelling. Nevertheless, baroclinic weather systems are fueled by contrast in thermal energy between two air masses, so descriptive terms for these are masses are still useful in subject hand analysis. That said, not every air mass fits cleanly into a specific category. The classifications are meant primarily to compare air masses across a region.

Air masses are traditionally classified according to the scheme developed by Tor Bergeron in 1928. This sorts air masses by moisture (continental, c; or maritime, m) and temperature (arctic, A; polar, P; tropical, T; or equatorial, E). The air mass may be further classified according to whether it is significantly warmer

FOG

HAZE

DRIZZLE

RAIN

RAIN SHOWERS

SNOW

THUNDERSTORM

Figure 5-3. Basic symbols for weather types. Variations are obtained by combining symbols. Multiple instances of precipitation symbols are drawn to indicate heavier occurrences (side by side, in triangle pattern, or in a diamond pattern).

or colder than the surface it is passing over. An air mass that is warmer than the Earth's surface gains the suffix "w". Since it is being cooled from below, it is prone to forming a stable inversion near the surface, favoring haze and leading to stratus and fog if it has high moisture content. Likewise an air mass that is cooler than the Earth's surface has the suffix "k", which means it is being heated from below. A "k" air mass is destabilizing, and as a result it is prone to wind gusts, low-level wind shear, and cumuliform clouds.

Once an air mass is identified, its abbreviation may be written near center of the air mass. The abbreviation is comprised of the moisture letter, the temperature letter, and the relative temperature letter if applicable. This annotation communicates basic information about the air mass to other forecasters and end users. Again it must be remembered that these classifications are idealized air mass archetypes, and an air mass may originate from other regions and have ambiguous characteristics. The primary archetypes are described as follows.

Figure 5-4. Analysis from January 1949 which includes air mass abbreviations: the cP mark in Texas and a mT air mass over cP air in Illinois. NOAA began phasing out air mass markings on many of its charts in 1968.

5.2.1. CONTINENTAL POLAR (cP). Continental polar air forms due to radiational cooling of air over cold or frozen terrain. Its most common source in North America is far northern Canada and the Arctic basin. In the winter, source regions can extend south into the United States if fresh snow cover blankets a large region. When cP air reaches the warm waters of the Gulf of

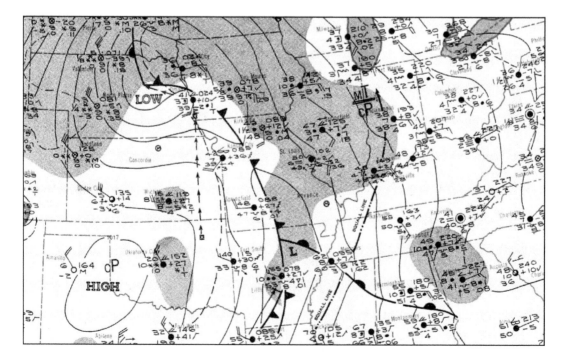

Figure 5-5. The secret to getting good at analysis is take charge of developing your skills and commit to doing it often. Every time a forecaster merely skims a surface chart during a significant weather episode, it's an opportunity that's forever lost. This binder contains all of the author's hand weather analysis charts spanning March 1, 1988 to May 3, 1989. *(Tim Vasquez)*

Sea-level pressure is expressed in an unusual fashion: in tens, units, and tenths of a millibar. This is a frequent source of confusion for beginning forecasters. The hundreds place is omitted but is always assumed to be 9 or 10 (use 10 if the figure is below "500"). Therefore a figure of "975" equals 997.5 mb and "063" equals 1006.3 mb. This "500" rule does not work around extremely deep lows or very strong highs and the forecaster may have to choose the appropriate value subjectively ("577" could represent 957.7 mb in low pressure or 1057.7 mb in high pressure). Needless to say this ambiguity has traditionally caused problems in weather map software.

Altimeter setting, if substituted for sea-level pressure, is also expressed in an unusual fashion: in units, tenths, and hundredths of an inch. The tens place is omitted but is always assumed to be 2 or 3 (use 3 if the figure is below "500"). Therefore a figure of "975" equals 29.75 inches and "063" equals 30.63 inches.

Mexico and Atlantic, it becomes cPk and forms extensive streets of stratocumulus and cumulus clouds.

5.2.2. CONTINENTAL TROPICAL (cT).

Continental tropical air is associated with strong heating of dry terrain by the sun. Its most common source region is Arizona, New Mexico, and the northern highlights of Mexico, especially during the spring and summer. The most common characteristics are fair skies, warm temperatures, and low dewpoints. The dryline, a surface feature found in the Great Plains, is a boundary separating this type of air from maritime tropical air, and where strong westerlies carry the light buoyant air eastward above the tropical air mass it is known as an elevated mixed layer (EML) and forms a capping inversion at the interface beneath. In the 1930s and 1940s, this EML was occasionally classified as an air mass type called Superior. The northerly incursion of maritime air in late summer causes the onset of "monsoon" rains in the Desert Southwest.

5.2.3. MARITIME POLAR (mP).

Maritime polar air forms when air masses stagnate over cool ocean surfaces. The temperature of an mP air mass is similar to that of the cool ocean waters it rests over, usually from 30 to 60 °F. Nearly all weather systems that come ashore on the west coast of the United States usher in maritime polar air.

Hadley's wind theory

"The same principle, as necessarily extends to the production of the west trade winds without the tropics. The air rarefied by the heat of the sun about the equatorial parts, being removed to make room for the air from the cooler parts, must rise upwards from the earth, and as it is a fluid, will then spread itself abroad over the other air, and so its motion in the upper region must be to the north and south from the Equator. Being got up at a distance from the surface of the earth, it will soon lose a great part of its heat, and thereby acquire density and gravity sufficient to make it approach its surface again, which may be supposed to be by the time it is arrived at those parts beyond the tropics where the westerly winds are found. Being supposed at first to have the velocity of the surface of the earth at the Equator, it will have a greater velocity than the parts it now arrives at, and thereby become a westerly wind, with strength proportional to the difference of velocity, which in several revolutions will be reduced to a certain degree, as is said before, of the easterly winds, at the Equator. And thus the air will continue to circulate, and gain and lose velocity by turns from the surface of the earth or sea, as it approaches to, or recedes from the equator."

GEORGE HADLEY

"Concerning the Cause of the General Trade Winds", 1735

5.2.4. MARITIME TROPICAL (mT). Maritime tropical air is found across all tropical oceans, and due to the high evaporation of these warm waters the air mass gains stout values of dewpoint — 60 to 80°F. Maritime tropical air contains abundant and rich moisture, but since the ocean absorbs most of the incoming solar energy, convective instability tends to be weak. When this air mass is advected onshore as a mTk air mass, it tends to gain considerably more heat and it destabilizes further, producing clouds and thunderstorms. Since tropical oceans cover a substantial percentage of the globe, maritime tropical air is the single most common air mass. Furthermore, the strong solar heating of land masses in the tropics allows extensive infiltration of this air inland. Even when the air mass moves inland, evapotranspiration from damp tropical terrain and vegetation helps offset the moisture lost from storm clouds and precipitation.

5.2.5. ARCTIC (A). Arctic air is generally considered to be somewhat different from continental polar air. In general usage, this applies to the harshest, coldest forms of continental polar air. However some forecasters prefer to apply this term to extremely cold air masses which are shallow (less than 1-2 km) and are undergoing intense radiational cooling. Interestingly, maritime arctic (mA) air does not exist, since by their very nature arctic environments are extremely dry.

5.2.6. EQUATORIAL (E). The equatorial air mass is a somewhat obscure term. It is considered to be an air mass near the equator with a modest temperature regime, and which is not affected by the subsidence and adiabatic warming found in the subtropical high. Therefore it tends to follow the intertropical convergence zone. Typical temperatures in the equatorial air mass are in the 70s and 80s.

5.3. Frontal concepts

A front is a boundary between two different air masses that have different temperatures. The frontal zone is the actual region of temperature contrast between two air masses, and is characterized by packing of isotherms or thickness lines.

5.3.1. FRONTAL LOCATION. The front always exists on the warm side of the frontal zone. In other words, the spot where the warm air begins showing a transition to cooler air marks the location of the front. Fronts almost always lie in pressure troughs and show cyclonic curvature of winds across it.

5.3.2. FRONTAL SURFACES. Fronts do extend upward into the atmosphere, sloping back toward the cooler air, and this upward extension of the front is referred to as the frontal surface. The surface front exists at the intersection of the frontal surface and the ground surface. An upper front exists at the intersection of the frontal surface and a given upper-level "surface" (such as the 850 mb level).

5.3.3. FRONTAL INVERSION. The frontal surface slopes back over cold air, so if a balloon is released in the cold air north of a front, it will soon pass into the warm air mass that exists aloft. As the balloon passes through the frontal zone it will show a reduction in the lapse rate or even warming with height. This is seen as an inversion on soundings, and it is called the frontal inversion. The frontal surface exists at the top of the frontal inversion. The dewpoint curve on the sounding usually shows an increase in dewpoint with height through the inversion, as opposed to radiational inversions where the dewpoint usually decreases through the inversion.

5.3.4. FRONTAL MOVEMENT. The front typically moves at a speed equalling the component of the wind flow across the front.

5.3.5. FRONTAL SLOPE. This term refers to the degree of slope of the frontal surface over the cold air. It is usually measured in terms of number of miles of run per mile of rise. For example, a frontal surface with 1:50 (1 mile of rise per 50 miles of run) slope is considered to be steep, while a frontal surface with a 1:300 slope is considered shallow. A steep slope indicates that the lift could be

Computerized processing of weather data greatly reduces the manual task of drawing isopleths. However a meteorologist cannot passively depend on computer output. Instead, he or she must master both drawing lines and improving automatically drawn weather charts. A meteorologist is expected to prepare charts better than the computer does, even if not as fast. For example, computer output is known for the absence of discontinuities (fronts), poor resolution of mesoscale processes, and other shortcomings. Therefore, a professional meteorologist must know the classical methods of analysis, including those developed by the Norwegian and Chicago schools and numerous other investigators who advanced our knowledge of the detailed structure of the atmosphere. Thus, studying graphical manual drawing of weather charts is not obsolete in the era of automation.

DUSAN DJURIC
"Weather Analysis", 1994

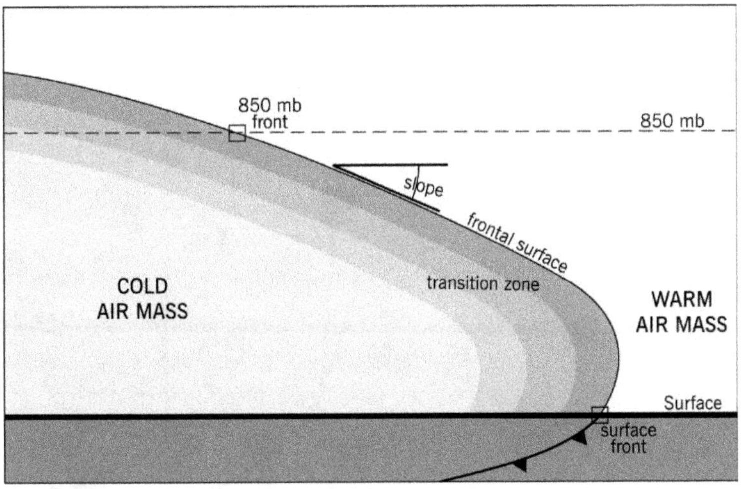

Figure 5-6. Cross section of a front with different areas labelled. The transition zone is where warm air transitions to cold, and it is physically part of the cold air mass.

strong if the wind flow forces the air to ascend the frontal surface. Frontal slope in the boundary layer is usually much steeper than that in the middle and upper atmosphere.

5.3.6. SECTORS. The warm air mass along a front is usually referred to as the "warm sector", while the cold air mass along a front is referred to as the "cold sector".

Figure 5-7. A shallow cold front in higher terrain stalling against the eastern slopes of the Rocky Mountains in Colorado and New Mexico on April 20, 2002. Albuquerque was reporting 72 degrees and fair skies, while the High Plains experienced cloudy weather and temperatures in the 40s and 50s. The cool, humid polar air undergoing upslope flow is largely responsible for the overcast stratus ceilings found across the High Plains. (Tim Vasquez)

5.3.7. FRONTOGENESIS. This term refers to the creation of a new front or the intensification of an existing front. This is reflected by an increase in the temperature gradient, an increase in thickness gradient, or strengthening of the frontal inversion. Frontogenesis is supported by low-level convergence in the wind field along its length, which tends to bring thermal contrasts together and makes them more intense. Frontogenesis can also be supported by diabatic processes (such as surface heating in the warm sector or surface cooling in the cold sector).

5.3.8. FRONTOLYSIS. This term refers to the dissipation of a front or the decrease in intensity of an existing front. This is reflected by a decrease in temperature gradient, a decrease in thickness

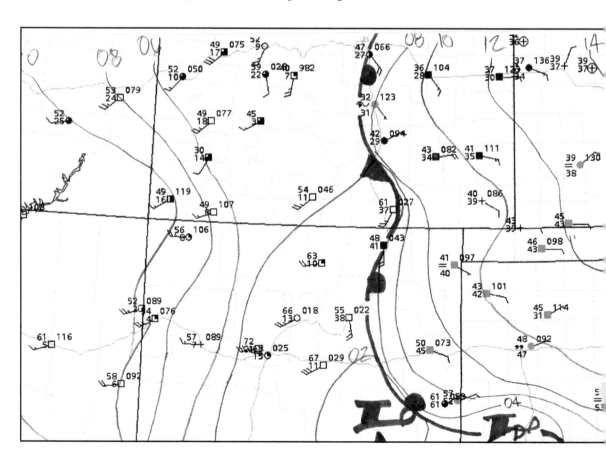

gradient, or decrease in the strength of the frontal inversion. Frontolysis is supported by low-level divergence in the wind field along its length, which tends to push apart existing thermal contrasts and make them less intense. Low-level divergence may be supported by upper-level dynamics. Frontolysis can also be supported by diabatic processes (such as surface heating in the cold sector or surface cooling in the warm sector).

5.4. Cold front

A cold front represents a front where a cold air mass is replacing a warmer air mass. Pressure tendencies tend to show a marked rise after the front passes. Forecasters must remember that like all fronts, a cold front is on the *warm side* of a transition zone. Therefore the arrival of a cold front at a weather station marks the instant that a transition to colder temperatures begins. Since a cold front is a function of temperature and not any other property, a front should never be placed along wind shift lines, troughs, cloud lines, and so forth if no transition zone exists. The boundary can still be marked but it should not be identified as a front.

5.4.1. ACTIVE COLD FRONT (ANAFRONT). This is a term used to describe a front where the warm air mass is being forced to ascend the frontal surface. It is associated with steep slope, uniform packing of thickness contours along its gradient, and a polar front jet axis which lies parallel to the surface front. Precipitation and clouds tend to exist behind the cold front.

5.4.2. INACTIVE COLD FRONT (KATAFRONT). This term describes a front where the warm air mass is being forced to descend the frontal surface. It is associated with a shallow slope, thickness contours which are not generally parallel to the front, and a polar front jet axis which is not parallel to the front. Precipitation, if any, tends to exist in convective lines ahead of the cold front. Skies are generally clear behind the front.

5.5. Warm front

A warm front is a front that ushers in a warm air mass replacing a cold air mass. Since fronts are found on the warm side of a transition zone, temperatures will rise gradually as the front approaches, then after its passage will rise less slowly or not at all. Pressure tendencies tend to stop falling or rise slowly after the

Wind shifts and fronts
Lines of wind shift with no proximity to bands of strong temperature contrast, moreover, appear relatively often on surface charts. The origins of such lines are not typically well-known, and they may arise from more than one source. The widespread practice of analyzing fronts along such wind shift lines is not appropriate. Such lines, including prefrontal wind shifts, should be denoted in some manner to distinguish them from true fronts and other surface boundaries.

CHUCK DOSWELL
"A Case for Detailed Surface Analysis," 1995

Blue norther
The temperature along the Texas Rio Grande had hit 100 deg F on the afternoon of February 3, 1899. However, as so often happens in late winter and early spring, the weather scene in Texas underwent a sudden turnabout. With nothing but strands of barbed wire separating the plains and prairies of Texas from the frozen tundra of the north polar region, a ponderous mass of glacial air poured through the state, forcing temperatures to unparalleled depths. By the time the Arctic air settled over Texas on the morning of Lincoln's birthday, the scales on some thermometers were nearly inadequate to gauge the intensity of the severe cold. Readings bottomed out below 0 deg F in virtually all of the northern two-thirds of Texas. The chill lowered the temperature to minus 23 deg F at Tulia in the southern Texas panhandle.

GEORGE W. BOMAR
"Texas Weather," 1983

front passes. The movement of a warm front is generally slower than that of a cold front.

Cold air masses tend to retreat slowly, and this makes warm fronts move slower than cold fronts. The low-level wind fields often have a strong southerly* component across the warm front. Isentropic surfaces always form a dome over cold air masses, and the sides of this dome are steepest within the transition zones. The wind blowing across a warm front is forced to ascend this dome in order to conserve its potential temperature.

Because the warm sector often contains considerable moisture and high relative humidities, this may quickly form layers of thick stratus, stratocumulus, and nimbostratus, and drizzle and fog may occur. If the warm air mass is convectively unstable, showers and thunderstorms may form. All of this occurs within the transition zone north of the front, and weather improves once the warm front passes.

Figure 5-8. Warm front in the Central Plains. Even without continuity to assess the front type, this is likely to be a warm front since the wind component across the front is generally south-to-north. *(21 March 2011 / 2200 UTC)*

5.6. Quasistationary front

An quasistationary front is the term for a front in which neither air mass is replacing another. The front may dissipate or

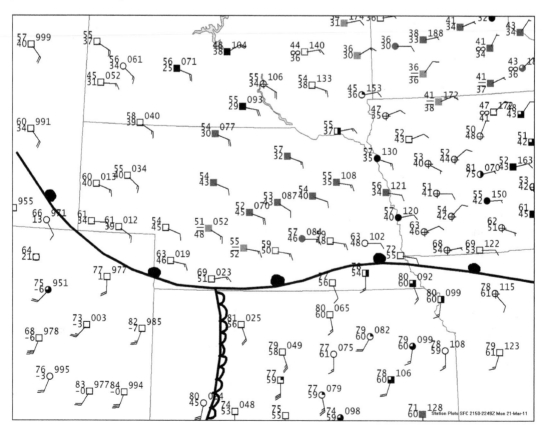

may begin moving again as a cold or warm front. Because of the slow or non-existent movement of a quasistationary front, they are generally considered to have the characteristics of a warm front.

Since this front is co-located with a transition zone and sloped isentropic surfaces, isentropic analysis and cross sections should be used to determine what kind of weather will occur along a quasistationary front. In short, low-level flow from cold to warm suggests descent, flow from warm to cold suggests ascent, and indeterminate flow suggests neutral vertical motion.

5.7. Occluded front

This is the term for a cold front which has merged with a warm front, bringing three distinct air masses into play. It is almost always caused by the fast-moving cold front overtaking a warm front near a baroclinic low. The air masses eventually are drawn into the surface low and begin combining, and they begin losing contrast on surface charts. Because the pressure trough is found where the deepest warm air is, isobars do not necessarily kink along the occluded front, and it does not necessarily lie in a trough.

5.7.1. COLD OCCLUSION. When the air behind the cold front is denser than the air ahead of the warm front, the cold front will undercut both air masses, forcing them aloft. The cold front

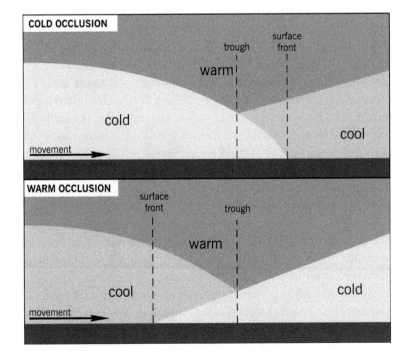

Figure 5-9. Occlusion types. The cold occlusion is the "classic" type, with the coldest air mass catching up to the cool air mass and continuing to behave like a cold front. The trough tends to shift somewhat behind the front near the deeper warm air. On the other hand, the warm occlusion occurs when the coldest air is associated with the warm front.

then becomes a cold occlusion. This is the most common type of occlusion. Since the cold front has undercut all air masses, the greatest depth of warm air aloft is *behind* the occluded front, so a pressure trough tends to occur behind the cold occlusion under the axis of deepest warm air.

5.7.2. WARM OCCLUSION. When the air behind the cold front is less dense than the air ahead of the warm front, the air mass behind the cold front will ascend over the warm front, leaving only the warm front at the surface. The warm front is then referred to as a warm occlusion. The greatest depth of warm air is *ahead* of the occluded front, so a pressure trough tends to occur ahead of a warm occlusion under the axis of deepest warm air. A warm occlusion is much more rare than a cold occlusion. One example where it occurs is along the Washington/British Columbia coast when a continental polar air mass has spilled westward into the Pacific. When a relatively mild Pacific weather system approaches this cold air, it is forced to ascend.

5.8. Dryline

Drylines demarcate a strong moisture gradient between warm tropical air and warm continental air. They are usually found in the Great Plains during the springtime, and may also be located in India and Australia. The dryline is significant because it is frequently associated with violent thunderstorm activity in the springtime.

5.8.1. STRUCTURE. The dryline is best visualized by looking at the moist air mass as a whole and thinking of the dryline as a boundary on the western edge. The moist air mass signifies the northward migration of tropical moisture. It intrudes into the dry continental air in the form of a shallow layer hugging the earth (anywhere from 1000-5000 ft deep or more). Therefore the dry air mass to the west of the dryline is pretty much identical to the air mass above. The dry air mass that exists above the moist sector is often referred to as the EML, or elevated mixed layer. Much of it originates from New Mexico and northern Mexico.

5.8.2. LOCATION. The dryline is located on the moist side of a moisture gradient. The moisture gradient itself is best found using dewpoint contours, or better yet, mixing ratio contours. The dryline is not a front, since it separates air masses with different moisture values rather than different densities. For example, temperatures in the dry sector can drop to 50 deg F at night and

From a veteran storm chaser:

I've driven through several drylines, and in most instances there is a sudden, recognizable change in air masses at the leading edge of the boundary. The most dramatic encounter I recall was while driving east from Lubbock, Texas when suddenly it seemed I hit a wall of water vapor. The parched air mass inside the car suddenly filled up with saturated air and the windows fogged up. Looking north and south along the dryline I was able to view the cross-section of two different air masses. A sharp clear sky was west of the dryline and distant cumulus were crisp and white. In contrast, the air mass to the east was hazy and cloudy, and I could see only a few rows of yellow, fuzzy cumulus fading into obscurity. Later that day, upon my return westward, I suddenly plunged through a wall of dust denoting the leading edge of the dryline. The muggy and oppressive air suddenly evacuated the vehicle and it felt like I had driven into a blast furnace. My lips became chapped and my teeth gritty.

TIM MARSHALL,
"Dryline Storms," 1988

rise to 100 deg F during the day, while the moist sector remains in the 70s and 80s. Therefore the dryline must not be located using temperature as criteria. A trough may be associated with the dryline; a cold front should not be drawn in such a trough unless it is clear that a dryline does not exist there.

5.8.3. SIGNIFICANCE. The dryline represents the westernmost extent of significant low-level moisture. Since upper-level dynamics move from west to east, having little effect in the dry sector, the dryline is the first location where thunderstorms develop. Also since lapse rates are steeper in the dry sector, momentum from upper-level winds tends to be transported downward, shifting winds throughout the dry sector, sometimes becoming gusty and raising dust, and adding to convergence of the wind field along the dryline. This convergence is often significant in producing upward vertical motion and severe weather along the dryline. Other small-scale circulations have been identified within the dryline that relate to moisture differences and contribute to convergence along it.

5.8.4. MOVEMENT. The dryline tends to move westward at night due to advection, as the wind flow brings higher amounts

Figure 5-10. Classic example of an active dryline as seen on 4 May 2003 in eastern Kansas. Three primary air masses are seen here — a cold polar air mass, a warm moist tropical air mass, and a dry air mass. The greatest risk of severe weather is near the triple point, that is, the intersection of the dryline and the front. *(Tim Vasquez)*

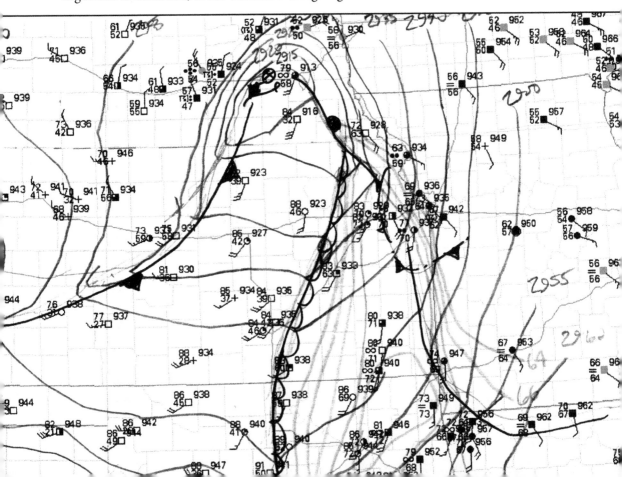

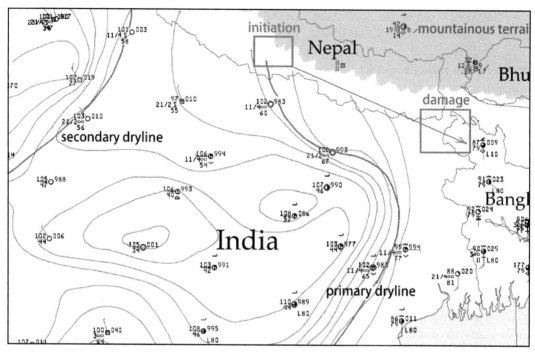

Figure 5-11. Drylines are not exclusively a Great Plains feature. This example from April 2010 shows a dryline associated with a damaging tornado in Bangladesh. Note the exceptionally high dewpoints in the moist sector. Tornadoes are a significant problem in this part of Asia and have largely been neglected from a research and emergency management standpoint. *(Tim Vasquez)*

of moisture to its western fringes. In these areas, the dewpoint will increase sharply at night and it will seem as if the dryline passed and moved westward. During the day, the dryline moves through mixing. As daytime heating progresses, convection tends to mix the moist air mass. Along the western fringes, much of it is dispersed into the dry layer aloft. A station in this area would note its dewpoint has fallen, and would infer that the dryline has passed and moved eastward. This daytime movement is often poorly related to the wind field and can only be predicted by examining the depth of the moist air mass and the amount of heating and mixing expected. This cyclic motion each day is sometimes referred to as "sloshing". The dryline is sometimes overtaken by a frontal system, especially late in the spring, and both the dry and moist sector are replaced with a polar air mass.

5.9. Outflow boundaries

Outflow boundaries are boundaries which separate a relatively undisturbed air mass from an air mass comprised of thunderstorm outflow. A fresh outflow boundary in proximity to a storm is known as a gust front, but even after the storm dissipates the boundary can persist for many hours or even days. It usually shows a subtle discontinuity of temperature or wind. Its importance in mesoscale forecasting has been emphasized in recent decades because not only can it serve as a zone where low-level convergence allows for thunderstorm development, but in certain cases it may contain sufficient shear to augment tornado production when ingested into the updraft.

Only the strongest, freshest outflow boundaries show up on synoptic-scale charts. The best tools for finding outflow boundaries are high-resolution visible satellite imagery, fine lines on radar, and mesoscale analyses. These boundaries should be continuously monitored as part of the forecast process.

5.10. Sea/land breeze fronts

These fronts are the result of unequal heating between the land and large bodies of water. They are most prominent in benign weather patterns, such as during the late summer. When strong weather systems are in the area, they can be displaced by the prevailing flow or suppressed altogether.

During the morning hours, the land heats the air above it at a much faster rate than that over the ocean, which causes it to rise. Air over the ocean then flows inland to compensate for the rising air. After a few hours, a circulation is established where air flows from sea to land at the surface and from land to sea aloft.

At nighttime, rapid cooling of the air over the land reverses the circulation, allowing air to flow from land to sea at the surface and from sea to land aloft. This circulation is much weaker than that of the sea breeze.

Figure 5-12. Streamline analysis can be valuable in finding minor boundaries, especially when pressure patterns are weak. This example is from May 15, 2003. *(Tim Vasquez)*

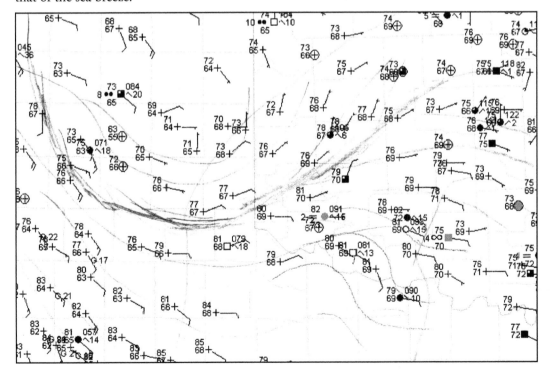

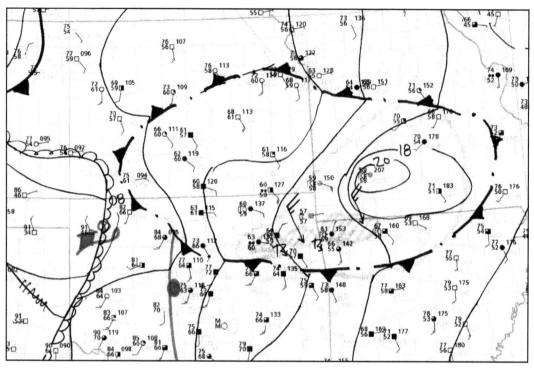

Figure 5-13. Large outflow area over Kansas and Missouri on 13 May 2003, produced by thunderstorms in extreme northwest Arkansas and northeast Oklahoma. The high pressure area within the outflow is often referred to as a "bubble high" or "mesohigh". Because of its higher density, the outflow air tends to undercut weak fronts and drylines like the ones shown to the southwest. *(Tim Vasquez)*

The sea breeze usually starts in the late morning hours, peaking in the mid-afternoon and ending before midnight. The corresponding "upper land breeze" lifespan is nearly the same as the sea breeze, and is usually at an altitude of 3000 to 9000 ft. At night, the land breeze is most established just before dawn, with the corresponding "upper sea breeze" occurring in mid-morning at an altitude of about 5000 to 7000 ft.

Chapter Five
REVIEW QUESTIONS

1. What might be a situation where an isodrosotherm analysis would prove to be valuable?

2. If a front has steep slope (the surface rises sharply with height), is the upper-level cold front likely to be found far away from the surface front or close to it?

3. Why is there an inversion in the transition zone on soundings?

4. Does convergence favor the strengthening of fronts or weakening, and why?

5. Where is the front located relative to a thermal gradient? Parallel to or perpendicular to it, and on which side?

6. What determines whether a front is a warm front or a cold front?

7. If a cold front with a 40°F air mass reaches a warm front with 20°F air ahead of it, what type of occlusion will result, and how does this occur?

8. What is the distinguishing characteristic of a dryline, as seen on surface charts?

9. What might be a reason that storms favor developing along the dryline?

10. Name some important techniques for finding outflow boundaries.

6 WEATHER SYSTEMS

Weather systems can be divided into two main classes: one where temperature advection is significant, such as in the frontal system, and one where it is not, such as in the subtropical high. If a system is driven by temperature advection, it is said to be baroclinic, and if not, it is said to be barotropic (equivalent barotropic).

Barotropic weather systems are not driven by thermal advection, so temperature gradients are typically weak. As a result, a jet stream usually does not directly overlie a barotropic system. Since any isotherms are in phase with the isobar field, no thermal advection is indicated in a barotropic system.

The baroclinic weather system is driven by baroclinic instability: horizontal differences in temperature. Such thermal contrast is unstable because the cold air has a tendency to sink and spread out beneath the warm air, and in turn the warm air moves to occupy the region formerly occupied by the cold air. This release of kinetic energy may escalate and produce a major circulation. When enough kinetic energy is depleted, the system dissipates and the thermal gradient has usually been mixed out. So in effect the gradient has been relieved. Naturally, all baroclinic weather systems have temperature advection as their primary feature. The isobars and isotherms also tend to be out of phase, and the system is normally associated with a jet stream over its center.

6.1. Baroclinic lows

A baroclinic low, also known as an extratropical cyclone or a frontal low, can form along any stagnant polar front boundary. It is responsible for the vast majority of cold season precipitation in the temperate latitudes. The evolution of the baroclinic low was first recognized by Vilhelm Bjerknes in the late 1910s and his conceptual model of evolution, the Norwegian cyclone model, quickly formed the backbone of modern operational forecasting.

6.1.1. THERMAL GRADIENT. The ideal state that precedes baroclinic development is a quiescent zone containing a thermal gradient, in other words, an idle stationary front. The isobars and isotherms are in phase, and there is no thermal advection taking place. However, this zone is considered to be baroclinically unstable. If an upper-level disturbance approaches, the surface pressure changes which precede it will distort the isobaric field, causing the isobars to no longer be in phase with the thermal field. The system becomes baroclinic. This begins the stage of baroclinic development.

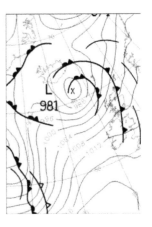

Title image
A large occlusion off the northwest coast of Ireland. This was a vertically stacked system with a 981 mb low at the surface. Note the presence of cold core convection, some of this possibly consisting of thunderstorms, due to the steep lapse rates associated with this sytem. The system appeared as shown above on the UK Met Office surface analysis. *(3 April 2011 / 1220 UTC)*

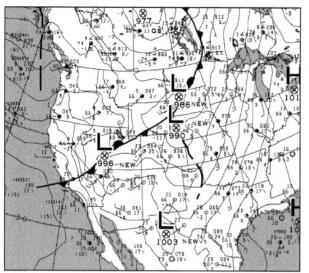 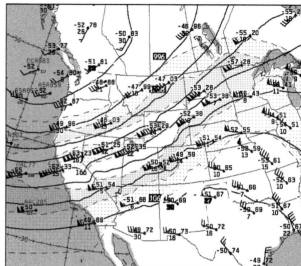

Figure 6-1. BAROCLINIC LOW. Baroclinic lows in western South Dakota and in Nevada. Each is directly below or slightly equatorward from a jet streak. (31 March 2010 / 0000 UTC; surface on left; 250 mb on right)

6.1.2. OPEN WAVE. Initially a low pressure area forms. This draws the cold and warm air masses closer together, intensifying the thermal contrast. This strengthens the upper-level features, and upper level charts may show intensification of the vorticity patterns and jet strength.

6.1.3. MATURE STAGE. Eventually the baroclinic low enters the mature stage. The deepening of the low is at its most intense during this stage. For this deepening to occur, there must be something removing mass from the system. The decrease in mass comes from a combination of upper-level divergence and the widespread release of latent heat as precipitation forms.

6.1.4. OCCLUSION. Gradually, the warm and cold fronts associated with the surface cyclone are forced to merge. This process cuts off the surface low from the warm sector. We see this on thickness charts as a movement of the surface low from the warm side of the thermal gradient to its middle, and eventually towards the cold pool. The occluded low begins filling, takes on barotropic characteristics, and eventually dissipates or absorbs itself into nearby semipermanent low pressure areas.

6.1.5. TRIPLE POINT CYCLOGENESIS. Though the occlusion is no longer a player, temperature contrasts are still very strong along the periphery of a warm sector. The triple point, at the cusp of

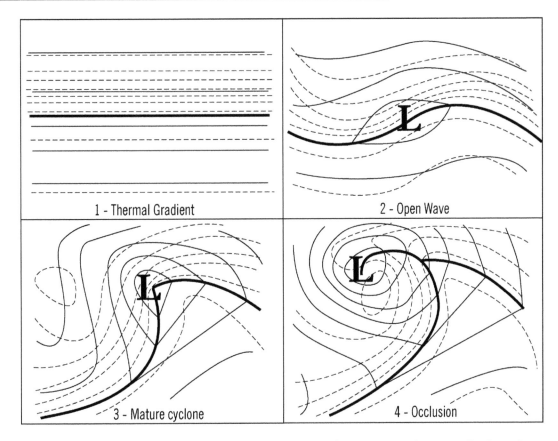

Figure 6-2. The four primary stages of baroclinic cyclogenesis. Surface isobars are drawn with a solid line, and low-level thickness isopleths are drawn with a dashed line. Frontal boundaries are depicted with a thick line. Note that the baroclinic low retreats on the cold side of the thermal gradient as it dissipates.

the warm sector, is a favored area for new cyclogenesis. This new cyclone matures and eventually follows the same occlusion stage as the original cyclone. A new triple point cyclone can form yet again.

6.1.6. EXPLOSIVE CYCLOGENESIS. Explosive cyclogenesis within a baroclinic low is often referred to as a "bomb". By definition, it occurs when the central pressure falls by more than 1 mb per hour for 24 hours. Exceptionally strong temperature contrasts are usually the catalyst for explosive cyclogenesis; a typical source is the zone where warm Gulf Stream waters meet the cold coastal waters. High lapse rates in the troposphere, strong upper level forcing, and rich low-level moisture are all contributing favors to rapid baroclinic development. The favored location for bombs are along the Atlantic Coast offshore from North Carolina, New England, the Canadian Maritimes, and Iceland. Where such storms graze New England, they are often referred to as Nor'easters.

6.1.7. CONVEYOR BELTS. During the 1970s there was a consensus among forecasters that extratropical cyclones were not comprised

Baroclinic low movement
Baroclinic low movement can be estimated by steering it with the 500 mb flow and moving it at 50% of the 500 mb wind speed.

Figure 6-3. **850 mb analysis showing a baroclinic system** about to exit the Rocky Mountain region. This brought a period of extremely cold weather to the Great Basin and Rocky Mountain region. The 700 mb and 850 mb charts make excellent tools for finding frontal systems in mountainous areas.

of a simple zone of cold air advection behind the cold front and warm advection ahead of the warm front. Cyclones showed evidence of distinctive circulations that extended into the vertical.

The most understandable example is the *warm conveyor belt* (WCB), in which parcels start near the surface in the warm sector, flow upward along the warm front surface, rise into the mid-troposphere, and eventually merge with the upper-level winds where they are swept eastward.

The *cold conveyor belt* (CCB) begins in the cold sector ahead of the warm front. Parcels are initially dry, but as they are swept westward into the cyclone, they receive moisture originating from precipitation falling out of the WCB. The CCB gradually ascends and reaches the area behind the surface low. The isentropic surfaces are sloped more steeply in the colder air feeding into the back of the system, which helps intensify the lift within the CCB. One branch of the CCB turns poleward (anticyclonically) as it rises to join with the upper-level winds, and another turns equatorward (cyclonically). This forms the deformation zone region of the baroclinic cloud system and contributes to precipitation on the rear side of the cyclone.

Finally the *dry conveyor belt* (DCB) begins in the upper troposphere behind the system. It descends in the cyclone's rear and is associated with clear skies, often forming a distinct dry slot.

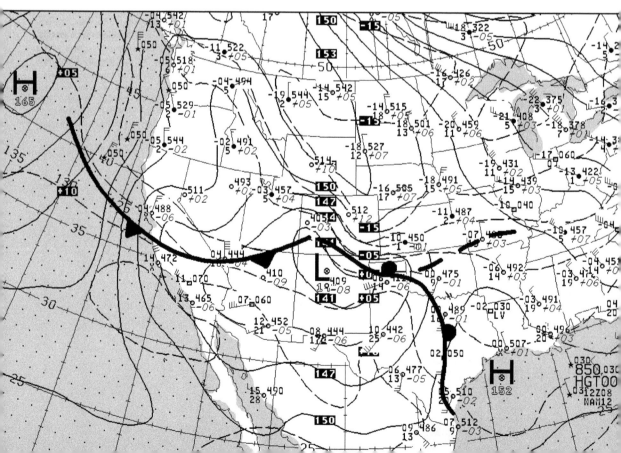

It should be emphasized that these are simplifications of the flow within the cyclone and may not necessarily fit every system. When appropriate, they describe the generalized airflow regimes within the system.

6.1.8. ELEVATED FRONTS. Some kinds of frontal systems do not adhere to the familiar Norwegian cyclone model, with the commonly expected signs of frontal passage. This usually occurs when upper-level conditions change much more quickly than the low-level portions, owing to terrain blockage, fast flow aloft, or other effects. When this occurs, the mid-level front "splits" from the low-level front, producing a so-called split front. In such a situation, cold front placement is ambiguous, warm fronts are difficult to find, and there is often a disconnect between precipitation fields and apparent frontal positions. The arrival of an elevated cold front and its associated cold advection aloft reduces the stability between low-level and mid-level air, and may result in development or enhancement of clouds and precipitation.

The arrival of an elevated cold front over a warm sector, with the cold front lagging at the surface, is most common in the Rocky Mountain region. The arrival of an elevated cold front over a cold sector tends to be associated with episodes of cold air damming. It is common in the Appalachian region during the winter season.

In either case, pressure tendency may be the best indicator of a frontal passage aloft, since surface pressure is related to density changes aloft.

6.1.9. POLAR LOWS. A polar low is an intense mesoscale cyclone that develops in a zone of cold air advection. It always contains deep convection, and stronger examples may develop spiral bands, cirrus outflow, and an eye, with the system appearing as a miniature hurricane on satellite imagery. Although the polar low contains a warm core, it is produced largely from occlusion of the core and subsidence rather than from latent heat release, and the system overall is baroclinic. The polar low forms along a prominent thermal gradient, such as the edge of an ice field or a zone where ice pack changes to open ocean. Some favored locations for polar lows are the far northeast Atlantic near Svalbard, in Hudson Bay, and in the Bering Sea. Generally the term *polar low* is used when the system is known to be producing winds of 35 kt or more. If not, it is referred to as a *polar mesoscale vortex*, not to be confused with the hemispheric *polar vortex*.

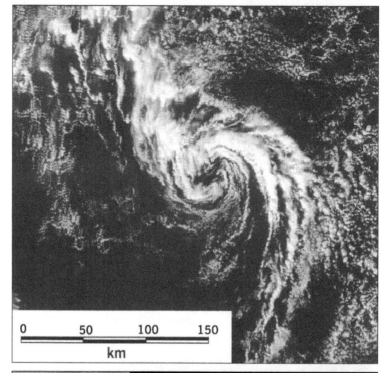

Figure 6-4. A weak polar low or polar mesoscale vortex in the Bering Sea near 59°N 173°E. This is viewed with 500 m Aqua imagery (top) and best available geostationary satellite image (bottom) using MTSAT-1R. Detection suffers greatly on the geostationary imagery due to foreshortening and the small scale of the system. The system measures no more than 100 miles in size, making this frame about the size of a traditional radar site image in the United States. *(29 March 2011 / 0040 UTC)*

6.2. Baroclinic high

A baroclinic high is in effect a "frontal high" that drives the cool air mass within a frontal system. A high pressure area cannot in itself contain a front, because the diverging surface air automatically weakens thermal contrasts. But baroclinic highs represent the source of air masses involved in frontal systems.

A cold air mass is initially barotropic because its temperature is homogenous. As a nearby wave develops, the air mass is drawn towards the baroclinic zone. A high pressure area forms due to the upper-level subsidence behind the wave.

Eventually, the air at the center warms due to subsidence and modification, and the air mass becomes barotropic again, being

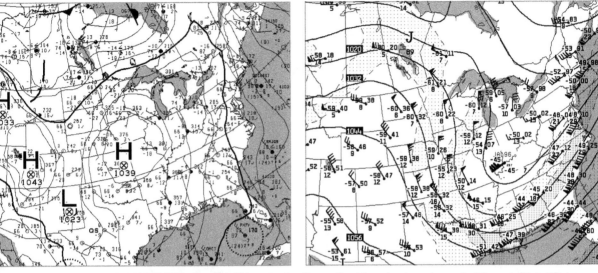

Figure 6-5. BAROCLINIC HIGH. Baroclinic high centered on Kansas City, Missouri and covering much of the central United States. Directly above it is a 65 kt upper-level jet. (10 January 2010 / 0000 UTC; surface on left; 250 mb on right)

absorbed into the subtropical ridge, often becoming part of the Bermuda high.

Braking mechanisms slow down and eventually stop the development of baroclinic highs. Although self-development causes baroclinic highs to strengthen, the intensification initiates other processes which slow down and ultimately stop the development. Braking mechanisms are much more efficient for highs than for lows, so highs rarely attain the same intensity as low pressure circulations.

As a baroclinic high builds due to convergence aloft, the low-level anticyclonic circulation increases. Anticyclonically-curved flow in the boundary layer causes low-level divergence, which will partially offset the mass being added by the system aloft. Friction slows down surface air parcels, which reduces the Coriolis effect and causes the pressure gradient force to be dominant; the winds diverge out of the high more strongly.

Subsidence is a warming process. To force this air to sink takes energy out of the high, reducing the energy available for development. It often limits the intensity that highs can attain.

When the 1000-500 mb thickness ribbon spreads apart within the high with time (a weakening thermal gradient), low-level divergence is predominating and the high is weakening. Another indicator of weakening is when the surface high center is located on the south side of the jet and has moved to a higher 1000-500 mb thickness line; adiabatic warming is indicated.

Baroclinic high movement
The movement of baroclinic highs can be estimated by taking 50% of the 500 mb flow or 70% of the 700 mb flow in the early development stages. Further development is favored when the short-wave ridge remains within 300 to 450 nm upstream. They may also strengthen under confluent flow in the jet stream pattern. When height rises increase over the surface high, this indicates that self-development is underway. Surface highs north of the jet or north of the tightest 1000-500 mb thickness gradient are favored for further development.

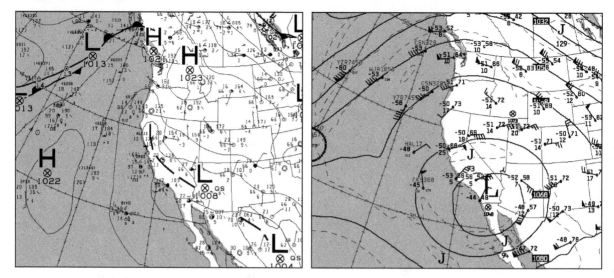

Figure 6-6. COLD-CORE BAROTROPIC LOW (CUTOFF LOW). The cutoff low becomes stronger with height. It is sometimes not apparent on the surface except for weak pressure falls and scattered shower activity. (20 October 2010; surface on left; 250 mb on right)

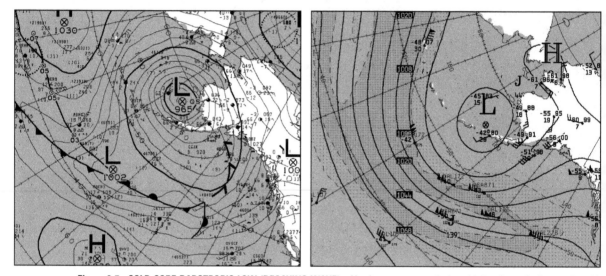

Figure 6-7. COLD-CORE BAROTROPIC LOW (DECAYING WAVE). Also known as an occlusion, this is a form of a barotropic cold-core low. The arrow markings east of the low mark a "trowal", the Canadian term for a "trough" of warm air aloft, signifying the axis of deepest tropical air above the occlusion. (17 October 2010; surface on left; 250 mb on right)

Finally, an indicator of weakening is when upper-level height rises are diminishing; upper-level support is decreasing.

6.3. Cold-core barotropic low

A cold-core barotropic low is comprised of cold air throughout the atmosphere. It causes low thicknesses within the air, resulting in low upper-level heights. Two examples are the occlusion, which is found poleward of the polar front jet, and the cutoff low, which is found equatorward. Their structures are both quite similar.

6.3.1. OCCLUSION. The most common example of a cold-core barotropic low results from the occlusion of a baroclinic system.

The occlusion usually fills due to boundary layer convergence. The cyclonic circulation strengthens with increasing height. The system is vertically stacked, but it may have a small amount of vertical tilt (i.e. an imaginary line drawn between the surface low and the 500 mb low will usually be vertical, but may tilt slightly from vertical). Since the coldest air is in the center of the low, the system is not directly associated with fronts. The occluded low is always the result of a baroclinic system which has occluded (the low has completely wrapped air masses into itself, which mix and

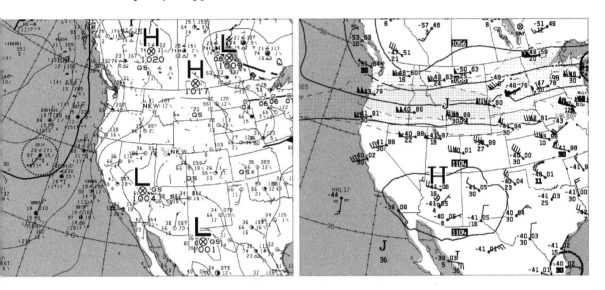

Figure 6-8. WARM-CORE BAROTROPIC LOW (HEAT LOW). Heat (thermal) low, which is a warm-core barotropic low. It is structurally related to the hurricane but gets its energy from insolation, rather than from the release of latent heat. Surface temperatures are as high as 42°C (108°F) in the region where California, Nevada, and Arizona converge. A heat low is found here at the surface and this translates to a weak co-located high aloft. (19 July 2010; surface on left; 250 mb on right)

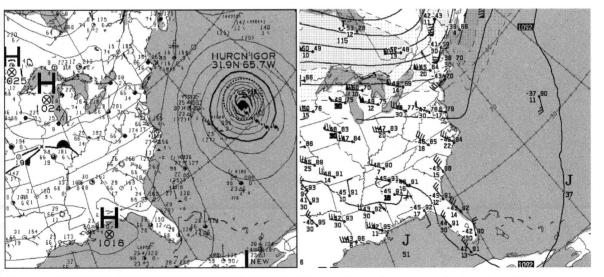

Figure 6-9. WARM-CORE BAROTROPIC LOW (TROPICAL CYCLONE). This type of feature is a warm-core barotropic low, which implies that high heights should be found aloft. The 500 mb chart, however, shows a low. It is frequently necessary to go to 300 or 200 mb to find the upper-level high over the storm. (20 September 2010; surface on left; 250 mb on right)

become homogenized). It occurs north of the polar front jet. The low usually weakens or absorbs into a larger semipermanent low pressure system, such as those found in the northern Atlantic or northern Pacific. While the occluded low is dissipating, fresh air mass contrasts to its south often lead to the development of new baroclinic system (new waves).

6.3.2. CUTOFF LOW. These are isolated upper-level lows that are found south of the polar front jet. They usually develop along the Pacific coast when a strong baroclinic high off of British Columbia transports cold polar air southward off the coast of Washington and California. The polar front jet stays to the north, causing the circulation to isolate itself and close off. The circulation is usually weak and difficult to find at the surface. Cutoff lows typically stay in the northwest Mexico region for days, then wander towards Texas or the Gulf of Mexico, where they may interact with rich moisture and produce thunderstorms. In rare cases a cutoff low may move westward into the Pacific and disappear. Most computer forecast models handle cutoff lows poorly; the most reliable movement indicator is Henry's Rule, which states that a strong short wave along the main polar front jet must come within 1000 miles of the cutoff low; the circulation will then open up.

6.4. Warm-core barotropic low

A warm-core barotropic low is comprised of warm air throughout the atmosphere. This warm air causes higher thicknesses within the air, resulting in high upper-level heights.

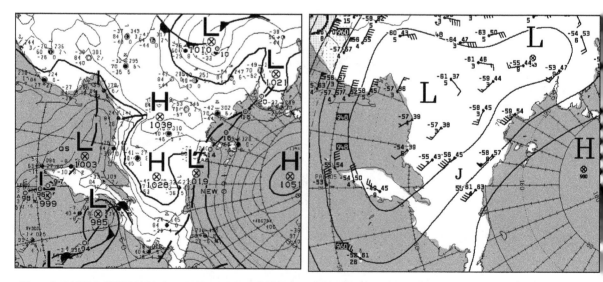

Figure 6-10. COLD-CORE BAROTROPIC HIGH (POLAR AIR MASS SOURCE). Cold-core barotropic high over northern Saskatchewan. Notice how it sits directly under a sharp trough aloft (low pressure). This high is continental polar air sitting in its source region over snow-covered areas of Canada. Two days later it evolved into a baroclinic high and shifted southward.

The low density of the warm air also causes low pressure at the surface. The system is barotropic, so it is vertically stacked, though it may have a slight amount of tilt. The cyclonic circulation weakens with increasing height, and may become a high aloft. Since the warmest air is in the center, no fronts are directly associated with the system. The low occurs south of the polar front jet, and is caused by low-level heating of the air mass. Any precipitation is relatively symmetrical around the center.

6.4.1. HEAT (THERMAL) LOW. The most obvious example of a warm-core barotropic low is the heat low, usually seen over the desert southwest during the summer months (a surface trough may extend into the San Joaquin Valley of California). Its energy source is diabatic, originating from strong solar heating on dry desert ground. The exact center of the surface low may be ambiguous and difficult to find, due to uneven heating. Diurnal thunderstorm activity may develop if there is sufficient moisture.

6.4.2. TROPICAL CYCLONE. Another example of a warm-core barotropic low is the tropical cyclone, which is related to the heat low but is more of an adiabatic system. It does receive substantial heat energy from the ocean surface, but the moisture present in the air mass and strong upward vertical motions cause extensive adiabatic warming due to the release of latent heat within the weather system. The circulation in a hurricane weakens only

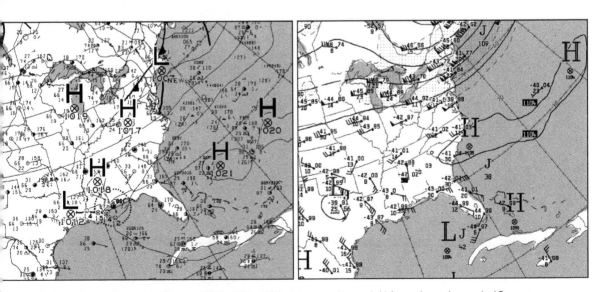

Figure 6-11. WARM-CORE BAROTROPIC HIGH (SUBTROPICAL HIGH). Warm-core barotropic high over the southeast united States. t translates to a high directly aloft. This is a classic example of a subtropical high, which usually serves as a source for maritime tropical air. (26 July 2010; surface on left; 250 mb on right)

gradually with height, and it is sometimes necessary to go up to 200 mb to find any anticyclonic circulation.

Some very intense extratropical cyclones, particularly bombs, may produce warm-core structures as they occlude. As a result, they may have structures similar to those of hurricanes and may even develop an "eye".

6.5. Cold-core barotropic high

The cold-core barotropic high contains cold air at the center of the high. The high density causes high surface pressures, while the low thicknesses result in low heights aloft. Since the system is barotropic, the system is vertically stacked, but may have a slight

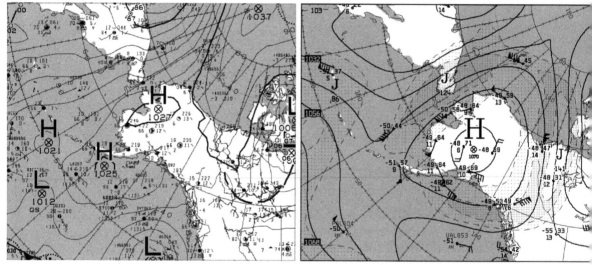

Figure 6-12. WARM-CORE BAROTROPIC HIGH (CUTOFF HIGH). A cutoff high, a form of warm-core barotropic high. It is very similar to a subtropical high. It is sometimes seen in Alaska due when strong insolation or latent heat release occurs north of the polar jet.

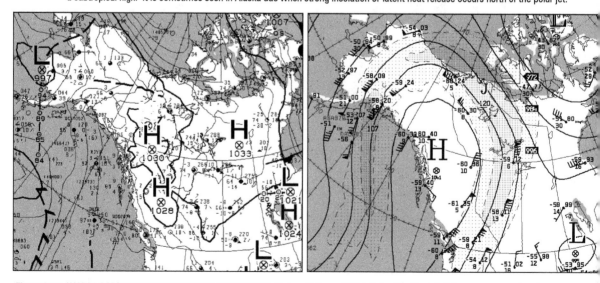

Figure 6-13. WARM-CORE BAROTROPIC HIGH (PLATEAU HIGH). Plateau high over British Columbia. There is little temperature advection around it, and an upper-level ridge sits above it. (20 February 2010; surface on left; 250 mb on right)

amount of tilt. The circulation weakens with increasing height. Since the coldest air is in the center, it is not directly associated with fronts. The high is usually north of the polar front jet.

The most common example of a cold-core barotropic high is a polar air mass in its source region. It forms due to intense cooling of the air in the low levels. As the air mass moves south, temperature contrasts in the air mass begin developing, causing the high to become a baroclinic high.

6.6. Warm-core barotropic high

The warm-core barotropic high contains a substantial amount of warm air throughout the atmosphere. Since it is barotropic, it is vertically stacked (but may have some tilt). The circulation strengthens with height. Since the warmest air is in the center of the system, it is not associated with fronts.

6.6.1. SUBTROPICAL HIGH. The subtropical high is a semipermanent feature in the subtropics. It tends to be strong both at the surface and aloft. The high is caused mainly by the upper-level poleward-moving current of air from the tropics, comprising the upper portion of the Hadley cell. The air tends to accumulate at about 30 deg latitude due to eastward deflection by the Coriolis force. This convergence causes an increase in mass and the air sinks (subsides). This produces a very broad area of light winds and relatively quiescent weather, though where sufficient moisture is available, cumuliform clouds and scattered areas of thunderstorms may be found.

The light winds within the subtropical high comprise a serious hazard to sailing ships, which may be left stranded for days or weeks. One popular legend holds that the region near 30° was named the "horse latitudes" by Spanish ships in subtropical highs which had to throw horses overboard to conserve resources. However, another theory cites the "dead horse ritual": an effigy of a horse was thrown overboard at the point where the ship's hands, having spent their advance pay in England and on the voyage, had finally worked off their debts. This usually occurred when the ships reached the subtropics.

6.6.2. CUTOFF HIGH. The cutoff high is a warm barotropic high located north of the polar front jet (either the main one or a southern branch of one). The surface high is typically weak. It forms when a warm air mass aloft is transported to a high latitude by a strong southerly flow. The flow later becomes more zonal, cutting off the upper-level high, leaving a warm pool to the north.

Cutoff highs are rare, but are seen more often in the Atlantic regions. When they do develop, they may cause an omega block pattern.

6.6.3. PLATEAU HIGH. The plateau high is very similar to the subtropical high, with relatively warm air aloft. However this warm air above the relatively cool ground establishes an inversion. Clouds and fog forms within the valleys, which in turn gradually radiates heat away at cloud top level, promoting cooling of the boundary layer. The increased air mass density within the boundary layer causes high pressure at the surface. This type of pattern sometimes occurs during the winter months in the northeast Great Basin region, leading to days or weeks of below-normal temperatures and fog.

6.7. Arctic air outbreaks

A very significant forecast problem in North America, and to a lesser extent, Europe, is caused by severe cold air outbreaks. These outbreaks also become a major player in winter storms. Large amounts of bitterly cold air can only be produced over large areas of snow cover. When more heat is radiated into space than is replaced from underneath by the Earth or overhead by sunlight, cooling occurs.

6.7.1. ARCTIC AIR MASS PRODUCTION. Snow cover on the Earth's surface is a key element in strong cold air mass formation. The snow cover insulates the atmosphere from the Earth itself. In the winter the soil tends to be relatively warm and even in the high arctic, it rarely falls very much below -20 to -30°C. Furthermore, snow is highly efficient at radiating heat away to outer space. The air near the surface conducts sensible heat to the snow, which then radiates the energy away.

Not only is snow cover important, but *fresh* snow cover is even more effective. Fresh snow contains substantial amounts of air trapped between the snow crystals, and air is an excellent insulator. But as the snow ages over a period of days and weeks, it settles and loses its composition of air, allowing more and more heat transfer through the snowpack.

All of this radiation results in exceptionally cold temperatures near the ground, with the top of the cooling layer marked by an inversion. This inversion is only centimeters deep when cooling begins but grows to a height of about 0.5 km over a period of days.

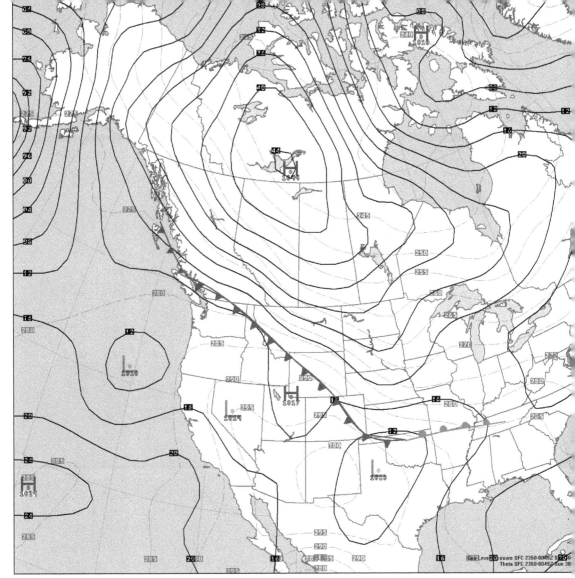

Figure 6-14. Classic appearance of a polar outbreak on the surface chart, showing the hallmark signals of anticyclogenesis in northwest Canada, intense ridging down the High Plains region, and a barrier effect along the Rockies. This outbreak in January-February 2011 was one of the strongest to appear in the Central Plains in ten years. *(30 January 2011 / 0000 UTC / Digital Atmosphere chart)*

6.7.2. THE ROLE OF CLOUDS AND FOG.

Radiation of heat by snow cover under clear skies is a key method for cooling an air mass. However it fails to explain the low temperatures in exceptionally cold air masses. This is ascribed to heat loss by fog, ice crystals, and cloud layers. Cooling increases relative humidity, and if the humidity is high enough, condensation will occur and these phenomena will develop.

For example, if the air temperature is cold but not severely cold, i.e. above -8°F (-22°C), and sufficient moisture is present, the initial radiational cooling will begin producing supercooled water droplets. This will result in either shallow freezing fog or low, thin stratus clouds. These layers of liquid droplets radiate considerable infrared radiation to the ground while blocking outgoing radiation, preventing surface temperatures from falling significantly. The top of the stratus or fog, however, radiates energy away to space and undergoes exceptionally strong cooling.

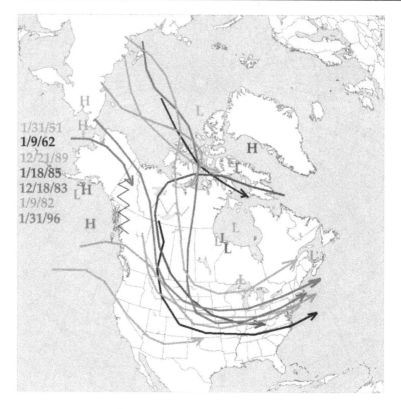

The net result is steady temperatures at the ground with fog and stratus building upward into the lower troposphere as cooling progresses.

High pressure is common and this may be accompanied by tropospheric subsidence. Downward motion will tend to produce adiabatic warming and transport drier air aloft toward the surface. This tends to offset the cooling somewhat and keeps freezing fog and ice crystal layers confined to shallow depths.

If the cooling is allowed to continue below about 0 to −10°F, this favors condensation directly into ice crystals: diamond dust. These are fine ice crystals that give the air a glittery appearance. Meanwhile the supercooled water gradually falls out. When the low levels contain ice crystals and are free of freezing fog and stratus, cooling then focuses entirely at the Earth's surface.

Freezing drizzle and light freezing rain events can be assessed by evaluating the cloud top temperatures (Huffman and Norman 1988). If the temperature is 0 to -10°C, this may be an indicator that precipitation layers consist mostly of supercooled water.

6.7.3. ANTICYCLOGENESIS. As cooling progresses over a large area, surface pressures rise and anticyclogenesis takes place. For large areas of cooling, the pressure rises are proportional to the amount of cooling that takes place. An exceptionally cold and deep air mass is usually required for the air mass to invade lower latitudes, so forecasters watch source regions carefully for signs of unusually strong cooling. The mass divergence at the surface or upper-level riding may result in subsidence, causing drier air to move from the mid-troposphere into the lower troposphere. This may diminish the amount of freezing fog and stratus, reducing the rate of cooling somewhat.

6.7.4. SOURCE REGIONS. In North America, the single most important source region for polar air is the Mackenzie River basin, an area centered on the southwest portion of Canada's Northwest Territories. Severe cold outbreaks are favored when the Aleutian Low is weak and when blocking patterns develop around the Alaskan region, a pattern associated with the –PNA teleconnection. This reduces the southerly wind component in northwest Canada, limiting the effects of downslope warming from the Rockies.

6.8. Winter weather systems

Winter weather brings a forecast challenge almost equal in complexity to that of severe thunderstorms. The stakes are high: rather than a few localities being raked by high winds or a tornado, an entire state may be buried in inches of ice or snow. Interestingly some of the techniques for forecasting winter weather have a close kinship with that of severe weather. The underlying processes are mesoscale in nature, and they require close examination of soundings, careful diagnosis of upper-level lift, and hour-by-hour analysis of surface charts.

In discussing winter weather, we will frequently refer to whether a layer or a sector is "warm" or "cold". This is shorthand for a temperature that is "above freezing" and "below freezing". The exception is when we discuss falling precipitation: melting and evaporation are a function of wet bulb temperature.

6.8.1. PRECIPITATION FORMATION. Winter weather is divided into four main precipitation types: rain, ice pellets (sleet), freezing rain, and snow. Rain is precipitation which is entirely in the liquid state, while freezing rain is perfectly identical to rain but simply freezes upon cold surfaces at ground level. Snow is an aggregation of ice crystals, while ice pellets consist of ice without crystalline shapes. All of the precipitation types can transition from one to the other depending on the layers encountered as they fall, however snow forms only within the ice crystal growth region. Precipitation that refreezes while falling will become sleet.

Rain is the most straightforward precipitation type. It usually forms in a warm layer. However, subfreezing air in the temperature range of 0 to –10°C may actually be dominated not

Ice crystal regimes

Type	T (deg C)
Thin plates	0 to -4
Columns, prisms, needles	-4 to -10
Thick plates	-10 to -12
Dendrites	-12 to -16
Sector plates	-16 to -22
Hollow columns, sheaths	< -22

by ice crystals by *supercooled water* . This means that even with a sounding that is completely cold, light freezing rain might occur. Rain may also be the result when the surface layer is sufficiently warm and deep to melt any type of precipitation that falls into it.

Freezing rain is identical to rain, but has two requirements: exposed surfaces which are below freezing, and a surface-based cold layer that is either fairly close to 0°C throughout its depth or is shallow. The latter prevents the rain drop from freezing, and the former ensures it does freeze upon contact. Because of this, the air temperature is not as important as the temperature of exposed surfaces. Even if air temperatures are 25°F, glaze from freezing rain will be limited to treetops, roofs, and bridges if the weather has been warm and sunny recently. For this reason forecasters should factor in soil temperatures and the character of recent weather.

Ice pellets occur when any type of precipitation changes from a liquid to a solid before reaching the ground. The key ingredient here is a layer of warm air aloft above a cold layer near the surface, consequently this pattern is usually only found in and around frontal systems or in cold areas capped by a low-level inversion.

Snow occurs when the cloud layer is sufficiently cold enough that it consists of ice crystals rather than supercooled droplets. This is likely when the cloud layer is colder than –10°C. Furthermore, development of snow at temperatures that favors dendrites (-12 to -16 °C) results in the highest efficiencies of snow production. This is because dendrite crystals are larger and tangle more efficiently with other ice crystals to produce snowflakes. This layer is known as the dendritic growth zone (DGZ) or dendritic growth layer. If strong ascent is forecast within the DGZ, then there is a high likelihood of heavy snow production.

6.8.2. ICE CRYSTAL GROWTH. Heterogeneous nucleation is the basic step in ice crystal formation within a snow cloud, particularly in the –10 to –20°C range. This occurs when droplets of supercooled water freeze directly on any available nuclei, such as dust and pollen. It produces an extremely small ice particle that may then grow further. Ice may also "materialize" directly onto nuclei, a process known as deposition. Both of these methods produce an extremely small ice particle less than 1 mm in size.

When ice crystals drift into warmer temperature layers, –10 to 0°C, supercooled water droplets freeze directly upon these crystals. This process is known as riming, or accretion. This produces large, fragile crystals. They frequently collide and splinter into pieces, providing new sets of condensation or riming nuclei. This is the ice multiplication process. This splintering greatly speeds ice crystal growth throughout the cloud and is the mechanism by which most snowflakes develop. However if too much riming occurs, graupel and sleet is the result. When all temperatures in a column above a station fall below -10 deg C, the ice multiplication process diminishes and this often results in a decrease in snowfall intensity.

If only one type of ice particle (dendrite, column, etc) is in a cloud, the particles tend to fall at similar speeds, avoiding colliding with one another. However when there are multiple types of ice particles, fall speeds are varied and the chance of ice crystal collision is greatly enhanced. This results in more rapid growth of ice crystals. For this to happen, a broad range of temperatures within the saturated layer (preferably the full range from 0 to -20 deg C) is best. Although temperature contrasts are important, if snowflakes fall into a deep, slightly-subfreezing layer with little temperature contrast, this will also cause snowflake growth. This is because snowflakes tend to be "sticky" at these warmer temperatures and this gives them time to aggregate. A 1000 ft depth of 0 to –5°C temperatures seems to be adequate for allowing this to happen.

Millions of crystals
Renowned cloud physics expert Vincent J. Schaefer estimated that it takes more than half a million ice crystals to cover a one-square-foot area with snow ten inches deep.

6.8.3. PRECIPITATION TYPE. Some simple rules of thumb exist for differentiating precipitation type. The most well-known rule is the 540 dam thickness line which establishes the "rain-snow line".

Figure 6-16. Snow depth is still measured with a ruler. Because of the rise of ASOS, observed snow depth data has become harder to get than in years past. Observer networks like COCORAHS are attempting to supplement this with reports from trained hobbyists. *(Tim Vasquez)*

Figure 6-17. Hexagonal dendrite snowflake as seen under an electron microscope. Snowflakes initially start in a growth region, where millions of microscopic ice crystals occupy a single cubic meter. They combine and grow into the familiar shapes seen here . *(USDA/BARC)*

Thickness and winter weather
Studies by Glahn 1975 and others showed that the 540 dam isopach often serves as the discriminator between liquid and solid precipitation near sea level. At higher elevations this changes to 546 dam at 3000 ft MSL, 552 dam at 6000 ft, and 564 dam at 9000 ft MSL. Cantin 1990 studied Canadian winter weather using other thickness levels and found a threshold value of 154 dam for the 850-700 mb thickness chart ("high layer") and a threshold zone of 129-131 dam for 1000-850 mb ("low layer"). Using these values, when the *high layer was cold* (thickness lower than the threshold value), a cold low layer favored snow and a warm low layer favored rain, while a transitional low layer allowed for liquid types if the high layer was marginally cold. When the *high layer was warm*, a warm low layer suggested rain, a cold low layer suggested mixed precipitation, and a transitional low layer suggested freezing rain.

Specific 850 mb isotherms have also been used with some degree of success. However these are based on an ideal winter weather system. Every weather situation is different and may contain very complex thermal structures in the vertical that are not accounted for by rules of thumb. The only reliable way for the forecaster to predict precipitation type is to develop a conceptual model of its life cycle from the growth stage to impact with the ground. This is called the *top-down method*.

First, the forecaster must determine which layer the precipitation will form within. This is bounded by the cloud top height on soundings and infrared satellite. Numerical models can help refine the levels where maximum ascent will occur. The temperature at this level dictates which type of precipitation particle will form. This is checked against actual soundings or forecast soundings to determine temperatures within the precipitation layer. At temperatures of 0 to –10°C, supercooled droplets and ice crystals will form, while below –10°C it will consist mostly of ice crystals and any supercooled water will tend to freeze into sleet before leaving the precipitation layer.

From there, the forecaster examines the temperatures that the precipitation will encounter while it falls. The question to ask is, *"Will any liquid precipitation encounter a cold layer, or will any solid precipitation encounter a warm layer?"* If so, this will change the precipitation type if it is sufficiently deep (more than 500-1000 ft, as a rule of thumb). For this purpose, cold or warm layers should be assessed in terms of wet-bulb temperature. In a saturated air mass wet-bulb temperature and air temperature will be equivalent, but they can differ sharply in a dry air mass, making precipitation type forecasting more difficult in dry regions like the western U.S..

6.8.4. PRECIPITATION LOCATION. The numerical models, fortunately, excel at predicting the location of synoptic-scale ingredients. There are some well-known rules of thumb in use which help provide the first guess for precipitation location. The heaviest snowfall occurs about 150 nm left [NH] of the surface cyclone track, and is heaviest at the time that the cyclone is undergoing most rapid deepening.

6.8.5. THERMAL ADVECTION. The forecaster should not just think vertically. Thermal advection, the horizontal transport of warm air and cold air around a system, is a strong contributor to changes in precipitation type and is governed by the strength of the tropical and polar air masses and the intensity of the wind field. Due to the complexity of adiabatic and diabatic effects within a weather system, the forecaster should avoid relying on isotherm analysis to assess thermal advection. This is one area where trends on numerical models can help.

6.8.6. DIABATIC PROCESSES. When precipitation falls into a dry layer, it will initially produce virga or light precipitation at the surface. Most of the precipitation evaporates in the dry subcloud layer, where it can produce massive amounts of diabatic cooling. Ultimately an entire column of warm subcloud air can cool to a subfreezing temperature. If precipitation continues, this cooling continues until the air reaches its wet bulb temperature, a process sometimes called *wet bulbing*. It has been shown that only 0.38 inches of liquid equivalent precipitation can cause an air mass to cool by 8 F°! Diabatic cooling can also result from a phase change from solid to liquid, such as snow falls into an elevated warm layer and melts. This can erode the depth and strength of the warm layer, resulting in a change with time towards solid precipitation types.

6.8.7. SNOW DENSITY. Making things even more complicated is that even if a forecaster

A study of heavy snow in the central and eastern U.S. (Goree and Younkin 1966) examined the track of the surface low and found that heavy snow usually occurred about 150 nm to the left [NH] of the track, about 300 nm in advance of the low, and was most intense when the low was deepening. Likewise a study of 850 mb charts (Browne and Younkin 1970) showed heavy snow was favored 90 nm to the left [NH] of the 850 mb cyclone track, with the -5°C isotherm bisecting the area of heavy snow.

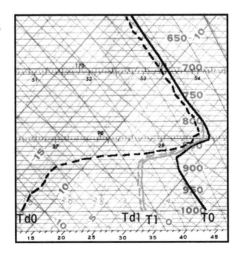

Figure 6-18. The effect of evaporative cooling on a sounding. The initial sounding is represented by the dark T0 and Td0 lines. Note the surface temperature of 4 deg C (39 deg F), the layer of warm air is at least 2000 ft deep (guaranteeing rain), and how the dry air occupies a layer between the surface and 875 mb. As precipitation falls into this dry air, evaporative cooling occurs. Some rain initially reaches the surface. Over the following hour or two, depending on the precipitation rate, the column cools to its wet bulb temperature. When "wet bulbing" is complete, the sounding profile is T1 and Td1. The entire column is subfreezing, snow is the precipitation type, and the surface temperature has cooled to -1 deg C (30 deg F).

correctly forecasts the intensity of the precipitation system, the snow depth forecast can be in error. Snow with a high snow-to-liquid ratio is fluffy, able to produce deep snow cover with a relatively small input of water, and is popularly called "dry snow". In the northern Plains where dry snow is common, ratios of 1:20 may occur, which means 20 inches of snow melts to 1 inch of water. Likewise, dense snow that produces substantial liquid amounts when melted is known as "wet snow", which involves ratios of about 1:8. A very cold column and production of unrimed dendrites and plates favors dry snow. Factors that favor wet snow include riming of crystals due to numerous supercooled droplets, snowfall through a marginally warm layer, and proximity to relatively warm ocean bodies.

6.8.8. WINTER WEATHER FORECASTING. Today's mesoscale models excel in outlining the temporal and spatial distribution of winter weather. They still, however, have considerable problems in anticipating precipitation intensity. Furthermore, winter weather episodes often develop unexpected mesoscale structures with small time and duration scales. Synoptic forecasting techniques like Q vector divergence and vorticity advection do not adequately account for these changes. The forecaster's job is to find these subsynoptic and mesoscale circulations. Analysis of the

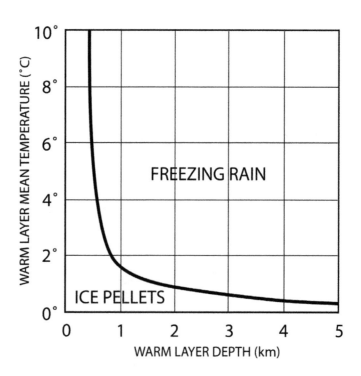

Figure 6-19. The Tau Technique (Cys et al 1996) is an empirical method that provides some guidance in difficult freezing rain vs. sleet situations. It requires the forecaster to measure the depth of the warm layer using soundings and other availabe tools and estimate the average temperature of the layer.

frontogenetic fields is one method commonly used to find localized areas of intense upward motion.

Though winter weather is not often associated with instability, steep mid-level lapse rates often occur in winter weather situations. Relatively unstable air within the precipitation layer tends to produce banded precipitation structures, while stable air produces stratiform structures.

It should also be pointed out that there are a vast number of empirical methods for winter weather forecasting, perhaps as many as there are for severe weather forecasting. These should all be used with caution, as rules of thumb are simplifications of elements found within very complex weather systems. If a rule of thumb has been successful with similar winter weather system structures in the past and the parameters it measures are appropriate for the type of pattern at hand, then it may be a good tool to use. Otherwise it should not form the basis for a forecast.

Chapter Six
REVIEW QUESTIONS

1. Describe the chain reaction of self-development with a baroclinic low.

2. Are coastal regions more favorable for baroclinic development during the summer or during the winter?

3. South of a baroclinic low, a subtropical jet begins producing extensive rain above the warm sector. The warm sector region has been very dry. How will this affect the baroclinic low?

4. How can you use thickness charts to evaluate where baroclinic low development is most likely?

5. What is the final stage in the life cycle of a baroclinic low?

6. Rain falls into a very thick layer of cold air (about 2500 feet deep), then strikes the ground, where the temperature is 25 deg F. What type of precipitation is occurring at the ground?

7. It is 30°F (0°C) and the sounding shows no inversions. What is the most likely precipitation type?

8. A warm front is south of Kansas City, which is currently reporting ice pellets (sleet). As the warm front approaches, what precipitation type transitions can be expected?

9. Why is heavy snow rare when the column of air above a station is below -10 degrees Celsius?

10. Ice pellets have been occurring all day. A television station's engineers need to climb a 1000 ft television tower to perform some urgent maintenance. Will the engineers most likely encounter snow or freezing rain?

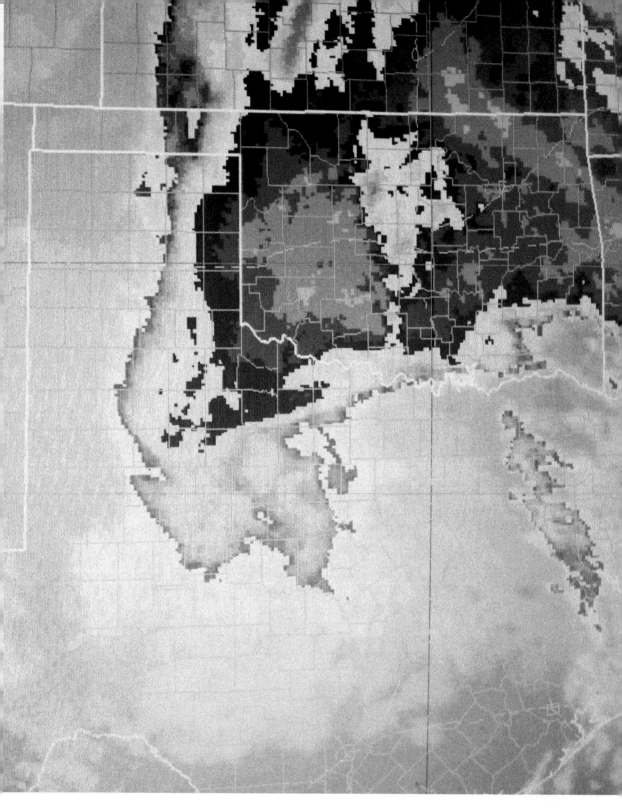

7 SATELLITE

No other modern meteorological invention has proved to be as revolutionary as the weather satellite. Not only can a satellite indicate cloud coverage across almost half of the Earth's surface, but it can analyze radiation signatures to yield information about cloud temperatures and water vapor, find thunderstorms using lightning detectors, and even relay weather information from central forecast agencies to users in the field. Its high temporal and spatial resolution also make it a vital element in the process of mesoscale forecasting.

7.1. Satellite orbits

The geostationary satellite is the most familiar to hobbyists. In this scheme, the satellite is launched and sent to a spot 22,236 statute miles (35,836 km) above the Equator, moving directly eastward at 6877 mph. If this is done, the satellite's movement precisely coincides with the earth's rotation. As a result, it appears to be stationary relative to any point on earth. The satellite must also be positioned so that it never drifts north or south of the Equator, otherwise the axis of its orbit would differ from the axis of the earth's rotation and the satellite would shift north and south each day. The satellite also cannot be brought closer to the earth since it would have to travel more slowly to remain above a given spot, and in doing so, it would fail to maintain orbit and would fall back to Earth.

In an entirely different class is the polar-orbiting satellite. Here there is no requirement to stay above a given position on earth, and the satellite is free to pass over all parts of the world. Generally such satellites have an orbit that is perpendicular to the equator, passing over the North and South poles during each orbit. Many agencies also prefer to put them in sun-synchronous orbit, where from the perspective of an observer on the sun, the satellite's orbit appears to be fixed, while the Earth rotates beneath. Any given spot can be observe twice a day. Such satellites orbit at about 600 miles above the Earth's surface and make one complete orbit in about 100 minutes.

The main disadvantage of polar orbiters is that a spot can only be viewed once during the day and once at night. Other problems historically have been that the satellite signal can only be received by line-of-sight, which at an orbit of 600 miles means that only end-users with special equipment have been able to receive the data. Furthermore, once the image has been received, it must be georeferenced, i.e. merged with geographic map outlines, which is no trivial task due to the complexity of the satellite orbit. As a result, their use diminished during the 1980s and 1990s as more

Title image
Snowstorm moving through the central U.S., as seen on a GEMPAK satellite display. (Tim Vasquez)

Figure 7-1. Warm lakes stand out on infrared imagery on unusually cold nights, giving a mottled appearance. When a lake suddenly disappears, this is a sign of cloud cover. It is also one way of differentiating fog from stratus. The wide streaks in the lower part of the image are cirrus clouds.

powerful geostationary satellites came online and provided more coverage of the globe.

The table has turned somewhat in the 2000s with the advent of a new generation of polar orbiters, particularly the NASA Aqua and Terra satellites. These contain new and advanced sensor packages that have made a significant impact on wildfire, lightning, and haze detection. The satellites provide a resolution of 250 meters, offering an incredible perspective on small-scale features such as dormant boundaries and cloud fields. Also significant advances in polar orbiter data acquisition means that data worldwide can be centrally collected and processed, instead of depending on end users to do the signal capture and processing.

7.2. Imagery types

The 1980s blockbuster movie *Predator* provided a dramatic demonstration to the general public that light is not needed to visualize a scene. All things are teeming with electromagnetic energy, and visible light is only one small part. Other electromagnetic "bands" exist: ultraviolet, X-ray, infrared, microwave, and radio. All these bands are the same thing; the only difference is the frequency at which the electromagnetic energy oscillates.

Weather satellites carry sensors that "view" the weather in different electromagnetic bands, However from a technical standpoint, only the visible and infrared spectrum are useful. Fortunately, the availability of infrared information in a scene of the Earth means that clouds can be monitored 24 hours a day, whether daylight is present or not.

7.2.1. VISIBLE. The visible channel measures light using the same wavelengths as the human eye, at an electromagnetic wavelength of about 0.5 microns. High amounts of brightness are caused by objects with high albedo, with albedo being a measure of the reflecting power of a surface. The largest albedo values are caused by cumulonimbus cloud tops and fresh snow. By contrast, forests and water surfaces produce low albedo. While visible imagery is naturally easy to use, its primary limitation is that it is only available during daylight hours. The exception is the polar orbiters under the umbrella of the Defense Meteorological Satellite Program (DMSP), which do carry low-light sensors and can provide coarse visible images at nighttime, its greatest benefit being for finding the location and extent of fog.

7.2.2. INFRARED. Infrared imagery in weather forecasting is traditionally measured at a wavelength of 12.8 microns, a wavelength known as far infrared. Imagery may also be sampled in the near infrared band, which is in the range of about 2 microns. Brightness is heavily a function of not albedo but temperature. This brightness scale is normally inverted in meteorological use so that warm land and water surfaces seem dark while polar regions and clouds seem bright. The result can be deceptive. A stratus layer might have a vastly different

Figure 7-2. Cold core convection around an upper-level low in southern California. The analysis chart shows heights and vorticity, while the imagery shows cumulonimbus clouds in an unstable air mass with steep lapse rates. Some of the storms are orographic, developing along the Coast Range. *(26 February 2011 / 2100 UTC)*

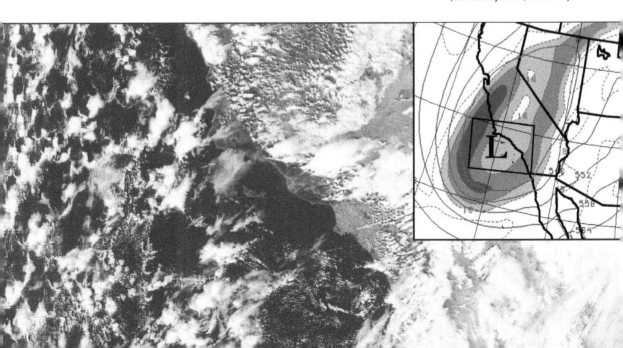

albedo than the forested region it covers, but on infrared imagery they might look the same due to the stratus and forest having the same temperature. Likewise, it might be impossible to detect a cold cirrus layer moving across a cold, snow-covered ground. Forecasters must be alert to situations like this and monitor trends in imagery for signatures like lakes and river patterns "disappearing" when clouds move over them.

7.2.3. WATER VAPOR. This channel measures the amount of radiation emitted by an object at around 6.7 microns. Water vapor heavily absorbs radiation at 6.7 microns, and absorption is greatest in the mid/upper levels of the troposphere, between 600 and 250 mb (15,000 to 34,000 ft). Therefore, where patterns are white we can assume that radiation from the lower levels of the atmosphere are being absorbed by mid/upper level moisture, and thus, mid/upper level moisture is present. Where areas are dark we assume that low-level radiation has not been absorbed by mid/upper level moisture, and thus, the mid/upper levels are dry. Note that because absorption is greatest in the mid/upper levels, it is impossible to draw conclusions about low-level moisture. Water vapor imagery brightness is also a function of the low-level temperature (the source of the low-level radiation); in cold regions, little radiation is emitted and we cannot make conclusive interpretations about water vapor imagery.

7.3. Satellite imagery limitations

Understanding the limitations of weather satellite technology is essential to making sense of what the images show. Some of the problems involve the satellite itself, while other issues center on the images themselves.

7.3.1. SENSOR RESOLUTION. The resolution of typical weather satellite sensors are limited to about 1 to 4 km. This means that any cloud features and textures smaller than this size will not be fully sampled and the pixels will present a blended or averaged appearance of all features within that area.

7.3.2. FORESHORTENING. While the satellite has a good view of locations underneath it is not able to accurately resolve cloud features low on the horizon. This means that geostationary satellites only obtain degraded resolution of polar regions. In many cases these areas must be supplemented with polar orbiter coverage.

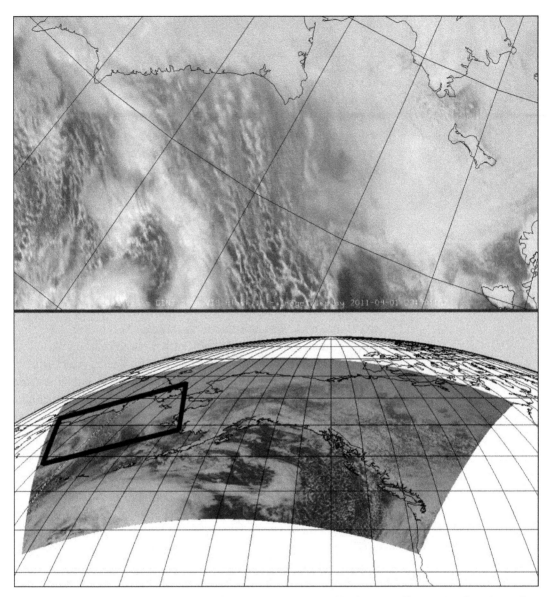

7.3.3. ATTENUATION. Dense haze will tend to reduce the albedo of low clouds, making them less detectable on visible imagery. In the same manner, moisture causes attenuation in infrared imagery, making low clouds difficult to detect. This is especially true in the tropics, where moisture tends to be very rich and very deep. All attenuation is amplified in areas where foreshortening is occurring.

7.3.4. CONTAMINATION. If a cloud layer is thin, it may not block out the view of cloud layers underneath. With infrared imagery this causes the indicated temperature values of the two layers to average out, giving a false indication of the actual cloud top temperatures and potentially giving misleading information about cloud top heights.

Figure 7-3. Foreshortening is more difficult to recognize nowadays than in years past, as satellite images are extensively remapped instead of presented as-is. For example, the visible imagery (top) shows the best possible GOES-WEST image of the Siberian Chukotka coastline. Detail is poor. Presenting the image as captured by the satellite (bottom) reveals the low slant angle as seen from the satellite's perspective. Foreshortening is one reason that polar orbiters are valuable in the Arctic and Antarctic. *(1 April 2010 / 2330 UTC)*

7.3.5. ENHANCEMENT. Since the human eye can only resolve about 20 shades of gray and a satellite image can deliver hundreds of possible shades, satellite imagery may be processed with an enhancement scheme, which assigns various color bands to specific color shades. Visible imagery is never enhanced except to adjust the contrast.

Infrared imagery, however, is widely subject to enhancement. From the 1970s through the 1990s, a number of enhancement schemes standardized by NOAA/NESDIS were in use, in which each color band represented a specific infrared temperature. The MB curve was the most well known of these and was a key part of the original definition of a mesoscale convective complex.

Today's enhancement schemes are very arbitrary and tend to be a function of what looks best on a particular website rather than what represents a certain temperature. This means that enhancement schemes vary significantly between various agencies and departments and usually have no function except to clarify features on the image. In some cases, a poor enhancement scheme may impair interpretation of the image or may obscure the cloud patterns in a wash of colors.

7.4. Clouds

The forecaster should be able to classify any type of cloud material seen on the imagery, as this links it to the weather processes which are occurring on the charts. Both brightness and texture are important. For instance, flat, dark cloud material on infrared imagery suggests fog, stratus, or stratocumulus, while flat bright cloud material indicates overcast layers of middle and high cloud. On visible imagery, cloud material is always a bright white, so flat dark layers suggest either image contamination, attenuation effects, or low-albedo sources such as dust, while flat bright layers indicate overcast layers of middle and high cloud. Fuzzy textures indicate cirriform cloud types while sharply-delineated edges suggest low cloud forms.

7.4.1. FOG AND STRATUS. This types shows as a bright cloud with very little texture on visible imagery, and on infrared imagery gives a dull flat appearance, often merging with the terrain owing to similar temperatures. Although stratus can be almost invisible on infrared imagery, it may be quite noticeable when it moves over lakes and suddenly blocks out the warm spots. On visible imagery, fog follows the terrain, with a preference for valleys and other low-lying areas.

7.4.2. CUMULUS CLOUDS. Cumulus clouds are readily apparent on 1 km visible imagery, appearing as small, dotted fields of clouds. In an unstable environment, some of the individual cumulus clouds may become larger and much brighter, producing what is known as enhanced cumulus. This is a sign that cumulonimbus development may be imminent. Due to the limited resolution of 4 km infrared imagery, there may be little differentiation between the cumulus clouds and the spaces in between, causing the cloud to appear as a broad, semi-continuous cloud layer.

7.4.3. CUMULONIMBUS. Cumulonimbus are striking features on all satellite imagery. On visible imagery they appear as bright, elliptical mesoscale spots, and may show overshooting tops where very bright pixels are located adjacent to darker pixels in the shadow area. On infrared imagery, cumulonimbus clouds appear as very cold, elliptical features. The overshooting top is easily visible where enhanced imagery shows a very cold spot. On severe storms, an "enhanced V" may form where very cold temperatures

Figure 7-4. Supercells in the Soviet Union on 9 June 1984 as seen with visible imagery. The storms here produced the disastrous Ivanovo and Yaroslavl tornadoes, killing 400. Censorship by the country's TASS news agency largely suppressed the story. Because of the satellite's position and the region's distant northerly location, the satellite appears only about 19 degrees above the horizon, introducing severe foreshortening artifacts and causing the coarse, striped textures seen here. *(METEOSAT / mapped by Tim Vasquez with McIDAS)*

seem to flow around the region downwind from the overshooting top, forming a V pattern.

7.4.4. STRATIFORM LAYERS. Thick stratiform layers include nimbostratus, altostratus, and dense cirrostratus layers. Because of the limited resolution of satellite imagery, it may also include overcast layers of stratocumulus or altocumulus that are sufficiently thick. These clouds appear as an extensive bright white sheet on visible imagery with a somewhat lumpy texture. On infrared imagery they appear as a uniform gray or white sheet. In baroclinic systems, multiple cloud sheets with sharply different temperatures may be present, indicating the presence of multiple layers.

7.4.5. CIRRIFORM LAYERS. On visible imagery, cirrus clouds have the same distinctive fibrous pattern as they do when observed from the ground. Thin cirriform clouds suffer from the effects of contamination and may not be detectable. On infrared imagery, their appearance depends a lot on the thickness and coverage of the layer; it may blend with clouds or terrain underneath or may appear as a cold, striated cloud layer with an extent of hundreds of miles.

7.5. Patterns

The forecaster assesses not only the cloud types present but also looks for distinctive patterns or shapes which indicate important meteorological processes. Some of the more important ones are described here.

7.5.1. WIND PATTERNS. The cloud formations may readily indicate the direction of the flow at that level, helping to fill in the surface or upper level analysis. This is especially true of cumulus cloud streets, whose axes are oriented parallel with the wind flow. Cirrus and most stratiform layers may also show evidence of alignment with the wind, showing fibrous streaks elongated with the wind. Forecasters should be careful, however, since what looks like streets or streaks may actually be a transverse band perpendicular to the streamline. Such transverse bands are most likely to occur where the layer is stable and contains vertical wind shear (a change in

Figure 7-5. Ordinary cumulonimbus clouds develop across west Texas as the summer monsoon pattern edges westward. The large, bright rounded cloud forms are cumulonimbus clouds and their fibrous cirriform anvils, while nearly all of the smaller globular clouds are examples of "enhanced cumulus", a manifestation of towering cumulus clouds and immature cumulonimbus clouds. *(1 August 1982 / 2031 UTC)*

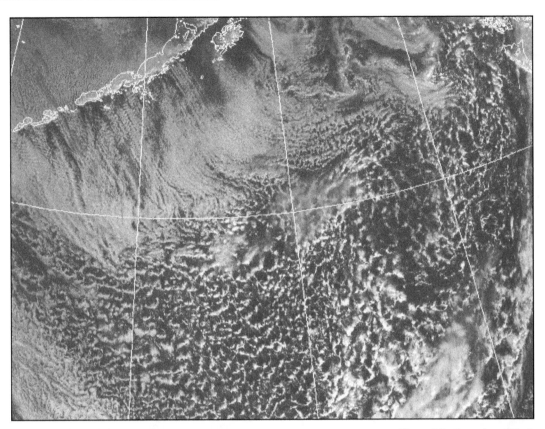

direction and speed with height) or is affected by gravity waves, such as from a mountain range upstream.

7.5.2. CLOUD CELLS. Layers of cumulus and stratocumulus which do not completely cover an area may exhibit a cellular appearance, with the individual cells usually measuring tens of miles in size. In an unstable environment, an open cell is the most common, where the clear spaces between clouds are interconnected. If the layer is stable, a closed cell is likely, where the clear spaces between clouds do not interconnect. In many cases, particularly over the ocean, the cell type hints at air mass modification effects: an air mass that is colder than the surface on which it lies is destabilizing and favors open-cell layers. This is common with cold air outbreaks and any air mass type suffixed with "k" (e.g. mPk and cPk). Likewise an air mass that is warmer than the surface below is cooling from below and stabilizing, favoring closed-cell cloud forms. This occurs in warm advection patterns and air masses with a "w" suffix (e.g. mPw, cPw).

7.5.3. FRONTS. Surface fronts are often a prominent feature on satellite imagery. When a baroclinic zone cloud system is not

Figure 7-6. Extensive cold air advection stratocumulus over the North Pacific south of the Alaskan Aleutian Islands on 13 February 2011 following strong advection of cold arctic air southward over the relatively warm ocean waters. Note the extensive open cell cloud forms throughout the bottom half of the image. *(NOAA/NWS Anchorage)*

involved, a surface front is best resolved with high-resolution visible imagery where a "rope cloud" is found or a line of showers. If a baroclinic system has developed but is only at the point of being a baroclinic leaf, the surface front is found along the equatorward side of the cloud formation.

In proximity to baroclinic zone cloud systems, the cold front lies along the tail end of the comma cloud, more toward the east [west] side of this tail with an anafront [katafront]. The warm front tends to be at a location that separates the comma tail from the baroclinic zone cirrus in the comma head. The triple point is where the fronts meet. If the system has progressed to the point of forming a prominent dry slot, it has likely begun the process of occlusion. The occluded front will be north of the jet along the back side of the low clouds in the comma head.

Figure 7-7. Though fronts and outflow boundaries are often delineated by rope clouds, sometimes a difference in cloud forms is the primary indicator. Here, the warm sector is south of the boundary, highlighted by cumulus cloud streets which are oriented perpendicular to the boundary. North of the boundary is stratocumulus formed into transverse waves, indicating a very stable air mass. *(25 March 2011 / 1745 UTC)*

7.5.4. LONG WAVES. The general pattern of cloudiness across a continent tends to define the long wave pattern. Where a large synoptic-scale area of clear weather transitions upstream to organized cirriform layers, a long wave ridge is indicated. These layers may show anticyclonic curvature, emphasizing the presence of the long wave ridge. Likewise, the transition from an organized weather system to relatively clear skies, particularly in the middle and upper troposphere, suggests the axis of a long wave trough. Upper level clouds in this region will show cyclonic curvature.

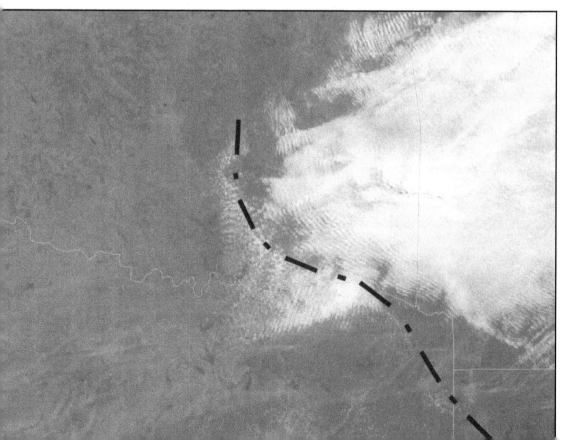

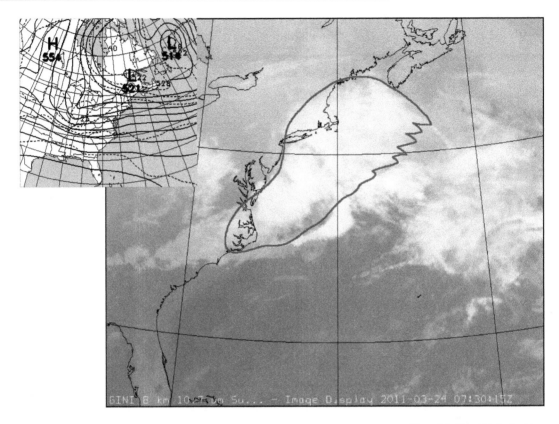

Figure 7-8. Baroclinic leaf pattern (outlined) as seen on infrared imagery. It represents a developing baroclinic system. The 500 mb analysis (left) showed no significant troughing in this area seven hours earlier, with the main trough axis in the American Midwest moving only to Ohio and West Virginia at the time of this image. *(24 March 2011 / 0730 UTC)*

7.5.5. Short Waves. Satellite imagery is likely to show short waves embedded in the upper-level flow. Short wave troughs are generally a mesoscale feature and either show no cloud signatures or are accompanied by relatively small areas of mid-level and upper-level clouds moving rapidly with the upper-level flow. The short wave trough axis is on the back side of this area of clouds. If the short wave trough moves across a cumulus field, as is common in the warm season, enhanced cumulus may develop along its leading edge, The short wave trough, where it encounters instability and low-level moisture, may generate enough large-scale lift to allow for the development of showers and thunderstorms.

7.5.6. Baroclinic Zone Cloud Systems. A baroclinic zone is any region which is strongly influenced by a temperature gradient. These regions may organize into large, multilayered cloud systems which have specific and recognizable structures.

The system initially starts as a baroclinic leaf, a very dense, elongated cloud that has a slight "s" shape on its upwind side. It is

Figure 7-9. The baroclinic leaf pattern is occasionally seen on satellite imagery. This is associated with the first stages of developing frontal systems. The upper-level jet position is usually as shown here, with a short wave extending along the back side of the system.

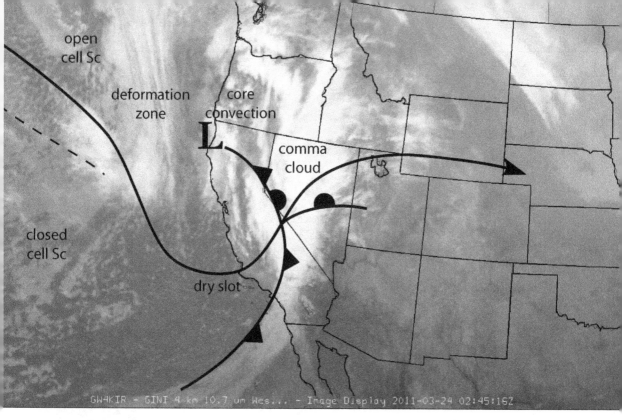

open
cell Sc

deformation
zone

core
convection

comma
cloud

closed
cell Sc

dry slot

GW4KIR - GINI 4 km 10.7 um Wes... - Image Display 2011-03-24 02:45:16Z

Figure 7-10. Classic example of a baroclinic system as seen on infrared satellite imagery. This image is unenhanced. The broad area of clouds over Nevada corresponds roughly to the location of the triple point and the upglide area north of the warm front. The primary surface low is found in proximity to the cyclonically-curved clouds in northwest California. As the deformation zone is well separated from the comma cloud, this is considered a "Type A" baroclinic cloud system; if not, it would be a "Type B" system. (3/24/2011 0245 UTC)

associated with an area of frontogenesis. As this system develops, it forms into a comma cloud, resembling the comma symbol. At this point, cyclogenesis has occurred along the front. This comma cloud is generally composed of nimbostratus, stratocumulus, and higher cloud layers, though in unstable conditions it may also consist of cumulonimbus clouds.

The upper-level low is usually found on the poleward side of the comma cloud. If the comma cloud is not well defined in this area and cloud layers are not very extensive, core convection may result. This occurs due to surface heating in the region of very steep lapse rates within the cold upper level low.

On the back side, a deformation zone cloud may develop poleward and to the rear of the comma cloud, and sometimes merged with it. This elongates along the axis of dilatation in the upper level flow and shrinks along the axis of contraction.

Behind the comma cloud, especially near the concave region on its upwind side, a prominent zone of clear air usually exists. This is known as a dry slot. It is associated with subsiding air on the back side of the system and low-level cold air advection.

7.5.7. SNOW COVER. Satellite imagery is perhaps the singularly most important tool for finding the extent of snow cover. However this always requires that the area not be obscured by cloud cover. Since fresh snow has extremely high albedo, visible imagery is the best type of image for detecting snow cover, though infrared imagery may depict the snow cover if the underlying ground temperature differs significantly. To the untrained eye, snow resembles an amorphous layer of low cloud or fog, but closer examination shows that it shows an irregular texture, owing to changes in albedo caused by terrain and vegetation. In mountainous areas this may give it a fractal- or veinlike appearance. Satellite animation is an exceptionally effective tool for discriminating snow cover from cloud layers.

7.5.8. JET STREAMS. Infrared imagery is useful for locating the polar front jet and subtropical jet. The poleward edge of cirrus shields tend to correspond to the axis of the polar or subtropical jet. The back side of baroclinic zone comma clouds, where the strongest concavity exists, corresponds to the jet stream axis. In the cold air advection zone behind baroclinic systems, if a transition between open cell and closed cell clouds are found, the jet axis lies about 50 to 200 nm poleward of this transition zone. The jet max itself has a tendency to be located in the dry slot of the baroclinic system.

Water vapor imagery is also helpful in locating the jet axis. Prominent areas of darkening behind baroclinic systems are the result of large-scale subsidence and drying. In straight-line flow, the jet axis is likely to be located in the center of this darkening, though with cyclonic [anticyclonic] flow it tends to be located poleward [equatorward] of the dark spot. Increased darkening of

Figure 7-11. Comma cloud. Baroclinic leafs often develop into large frontal systems manifesting theirselves on satellite imagery as "comma clouds". This conceptual model of a comma cloud shows the approximate location of surface fronts and the upper level jet.

Figure 7-12. Snow cover, leaving a telltale track of an eastward moving winter storm across the Great Lakes region. The snow cover shows highly irregular textures due to varying terrain and vegetation. Bright white spots are frozen lakes. Overlaid on the snow track are broad layers of stratocumulus and cumulus, formed up into cloud streets in the cold northerly flow. (24 March 2011 / 1515 UTC)

this area may indicate an increase in the strength of the associated jet max.

7.5.9. LEE CIRRUS. Cirrus bands may form on the lee sides of mountains in the presence of a strong component of flow across the mountain range and relatively high upper tropospheric humidity. In the United States, it is especially common in the High Plains region. The lee cirrus may contain transverse waves, which are ripples oriented perpendicular to the flow and are formed by the presence of gravity waves. This is usually more common with subtropical jet patterns. Lee cirrus formations may also be associated with standing lenticular altocumulus formations. As with one of the jet stream placement rules above, the jet axis is usually found along the northern edge of this lee cirrus zone.

7.5.10. DUST PLUMES. When high winds loft soil from dry agricultural areas, mesoscale dust plumes may occur. These appear

Figure 7-13. Duststorm in West Texas on 24 February 2007 as seen on visible satellite imagery, showing a massive dust plume (dull texture) originating near Lubbock and Midland, sweeping northwest. Gravity waves give the dust plume a somewhat rippled texture. A much brighter area of stratocumulus and scattered cirrus is at the top and middle of the image. The dust event led to airport closures at Dallas-Fort Worth, where it reduced visibilities to less than a mile.

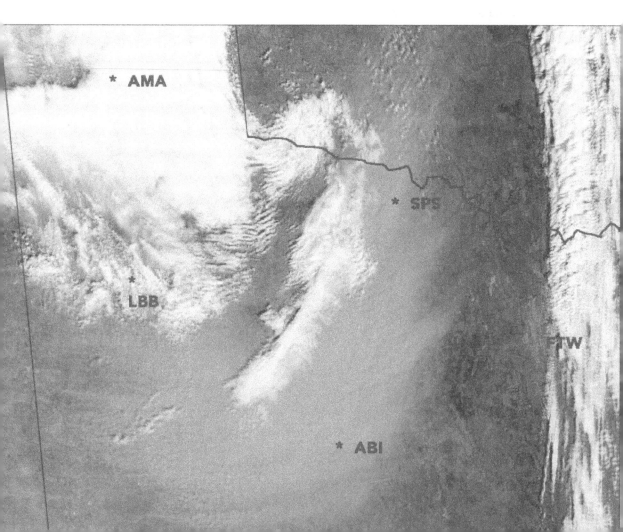

on visible imagery as diffuse areas of medium albedo, while on infrared imagery they are not easily detected.

Chapter Seven
REVIEW QUESTIONS

1. Why is polar orbiter imagery especially useful in polar regions?

2. What type of cloud has the highest albedo on satellite imagery?

3. Exactly what is happening to the radiation in the troposphere which causes a moist region in the tropics to show as "whitish", i.e. moist, on water vapor imagery?

4. What cloud forms would you expect when a very cold air mass moves over warm ocean waters?

5. Where is the jet stream located in relation to open cell and closed cell patterns on satellite imagery?

6. What does the comma cloud indicate on satellite imagery?

7. What is the concave area in the back of a comma cloud and what causes it?

8. Where is cloud material located relative to a short wave trough axis?

9. Explain how animation can be used to determine whether a pattern is stratus or snow cover.

10. Explain how lakes can be used to check for the presence of stratus layers at night.

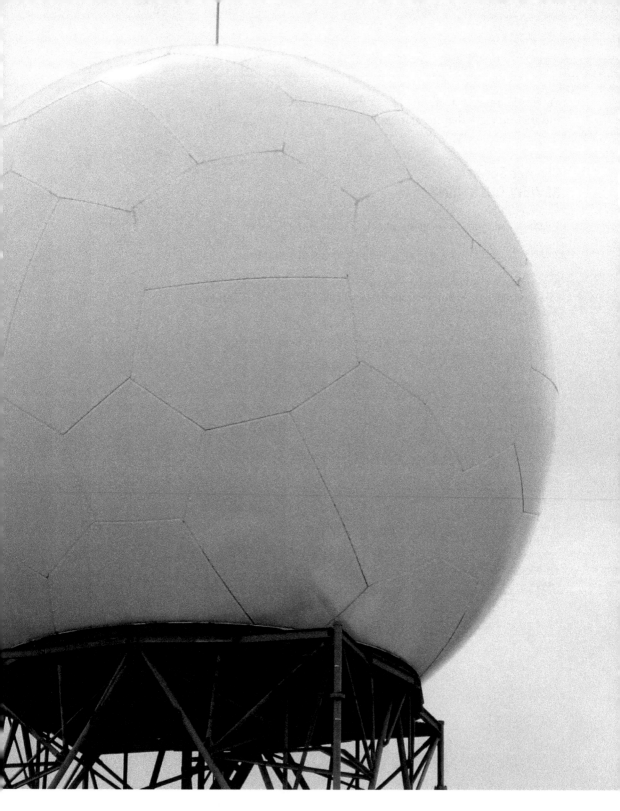

8 RADAR

Radar technology during World War II was primarily a means to defend against enemy forces, but its use quickly expanded to weather forecasting. Radar meteorology entered a new age in the 1990s when Doppler radar units finally spread beyond research circles and became the backbone of modern forecast agencies.

The United States radar network is now composed of over 100 WSR-88D (NEXRAD, or Next Generation Weather Radar) Doppler radar units. Images are distributed over weather datastreams and are widely available over the Internet. Not only does the United States have a truly modern radar network, but as of 2011 the country was rolling out a major technology upgrade known as dual polarization radar.

It should be emphasized that other countries such as Canada, Great Britain, China, Australia, and South Korea have recently fielded modern Doppler radar networks, but even as of 2011 the distribution of data in those countries is heavily restricted or is sold as a commercial service. As a result, most of this chapter is developed around the United States radar system, where public policy mandates open access to all official meteorological data. Many of the principles here apply, though, to all weather radars.

8.1. How radar works

Radar is a system that measures the speed of radio energy, which in turn moves at the speed of light. A powerful radio transmitter in the radar unit broadcasts a pulse of energy in a certain direction, listens for the reflection, measures the time elapsed, and converts this to a distance figure. With distance established, all that's left is to determine azimuth and elevation, and this can be obtained simply by checking where the dish antenna has been pointed.

Finding direction, distance, and elevation of an echo was suitable for World War II stations that simply needed to pinpoint where enemy aircraft were. But for meteorologists, a radar unit can actually be designed to analyze the echo's properties to find all sorts of characteristics about the atmosphere where the echo originated. Here we describe the three basic properties of reflectivity, velocity, and spectrum width. Advanced properties will be described in a separate section on dual polarization.

8.2. Reflectivity

Reflectivity (Z_H, or Z_{HH}) is a measure of the power, or intensity, of electromagnetic energy received by the radar. This

Considerable radar research was done in the 1930s and 1940s, leading to decisive turns of events during World War II. One of the first radar networks was England's Chain Home early warning system, which had no steerable dishes and could not determine azimuth, but could detect the telltale echoes from approaching German bombers. Ironically the Germans were able to detect the echoes and use them to accurately triangulate the location of British planes. By 1943 England had fielded several additional radar networks with higher quality and longer range, allowing the country to keep track of everything from German bombing raids to V-2 missile launches. Germany also took great initiative to develop its own radar network, starting with the slightly less capable Freya system in 1938. It was largely responsible for the loss of 14 British planes in December 1939 which were attacking a German naval base, and this caused the British forces to turn to night bombing tactics for the remainder of the war.

Title image
The Oklahoma City Twin Lakes WSR-88D radome, *(Tim Vasquez)*

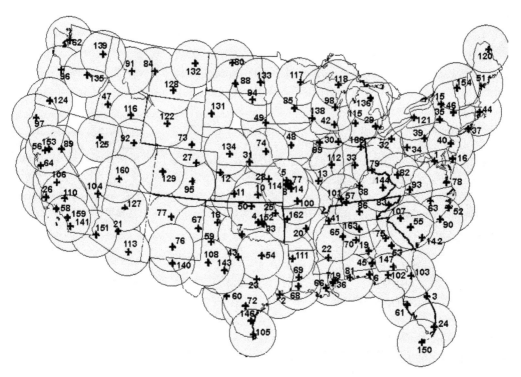

Figure 8-1. WSR-88D coverage in the conterminous United States as of October 2000. The digits indicate each site's sequence number. *(Courtesy NOAA/OSF)*

is expressed in decibels. The energy normally comes from water drops, drizzle, snowflakes, and hail. Weather radar is not designed to detect cloud droplets, however, since the drops are too small and cannot be detected reliably without switching to a much higher wavelength radar, which suffers from significant attenuation in rain events.

Therefore, conventional weather radars are optimized to detect larger particles like rain drops, snowflakes, and hail. Only the thickest of clouds where the early stages of precipitation are occurring will produce any detectable reflectivity values. Rainy precipitation areas and especially hail produce high values of reflectivity. The highest possible radar reflectivities come from wet hailstones.

8.3. Velocity

Velocity (V) on a Doppler radar indicates *outbound velocity* (thus, outbound velocity is positive and inbound velocity is negative). If a radar target is in motion relative to the radar, the reflected energy will come back at a slightly different frequency, a

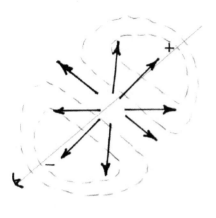

Figure 8-2a. Pure divergence shows a velocity couplet along the radar beam, with negative (inbound) flow on the near side of the circulation and positive (outbound) flow on the far side.

Figure 8-2b. Pure convergence shows a velocity couplet along the radar beam, with negative (inbound) flow on the far side of the circulation and positive (outbound) flow on the near side.

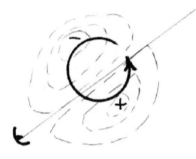

Figure 8-2c. Pure cyclonic rotation shows a velocity couplet across the radar beam, with negative (inbound) flow on the left side of the circulation and positive (outbound) flow on the right side.

Figure 8-2d. Pure anticyclonic rotation shows a velocity couplet across the radar beam, with negative (inbound) flow on the right side and positive (outbound) flow on the left side.

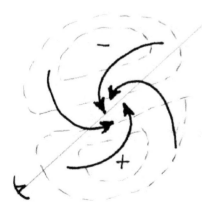

Figure 8-2e. Cyclonic convergence, as is often seen in severe storm mesocyclones, shows a couplet that is diagonal to the radar beam. Negative (inbound) flow is on the far left side of the beam and positive (outbound) flow is on the near right side.

Figure 8-2f. Anticyclonic convergence, which we might see in a rare anticyclonic tornado, shows a couplet that is diagonal to the radar beam. Negative (inbound) flow is on the far right side of the beam and positive (outbound) flow is on the near left side.

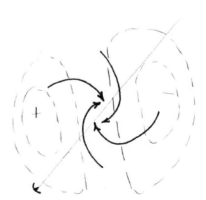

phenomenon called the Doppler effect. This frequency shift can be measured to obtain information about how fast the target is moving. The WSR-88D Doppler radar does not actually measure this frequency, however. It uses an algorithm called pulse-pair processing, where the wavetrains from two consecutive pulses are compared and the differences yield the velocity information. This algorithm is important for meteorologists to understand since it imposes limitations on the data in various weather situations. This will be covered later.

The key thing to remember about velocity is that only measurement to or from the radar is measured. It is not possible to detect movement perpendicular to a radar with a conventional Doppler radar. The full signature of rotation in a circulation like a tornado will not be seen; rather, only the portions moving most rapidly toward the radar unit and most rapidly away will stand out. This type of signature is known as a couplet, and its relation to the radar beam can be easily analyzed to find out whether it marks cyclonic rotation, anticyclonic rotation, divergence, convergence, or even a mix. See Figure 8-2 for an explanation of how couplets are analyzed.

8.4. Spectrum width

The WSR-88D radar is capable of measuring the variance of velocity that exists within a given sample volume. This is known as spectrum width (SW) and is given in decibels (dB). The correlation of spectrum width with weather phenomena has not been well-studied, so this product is rarely used. However it has been found to a significant marker for tornado debris clouds, where wind forces on objects with widely differing sizes and shapes produces an extremely diverse range of velocities.

8.5. Dual-polarization data

In 2011 and 2012, the nation's entire WSR-88D network will upgrade to dual-polarization technology. This examines the radar energy in two different planes: in the up-down and left-right directions. The radar still measures all the same base elements as on earlier radars: reflectivity, velocity, and spectrum width. However four primary data elements are added which compare the energy coming from each plane. By doing this, the radar can obtain valuable information about the shapes of the particles being detected, giving the ability to differentiate precipitation types in different parts of the cloud.

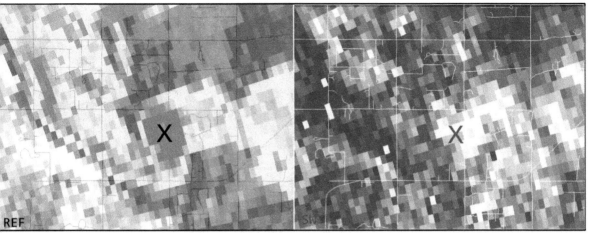

10 MAY 2010   2250 UTC   LITTLE AXE OK

8.5.1. DIFFERENTIAL REFLECTIVITY (Z_{DR}), dB. This measurement determines the ratio between power reflected in the horizontal plane from power reflected in the vertical plane. In short, Z_{DR} = 10 log (Z_h / Z_v). This formula shows that when horizontally-polarized reflectivity is dominant, ZDR is positive, and when vertically-polarized reflectivity is dominant, ZDR is negative. In terms of precipitation, it gives information on the shape of the drop. Large droplets and insects tend to have horizontally-oblate signatures with a Z_{DR} of +2 to +4, while snow, ice, and tornado debris tends to be close to 0. Hail and conical ice shapes have a negative Z_{DR}. A drop in Z_{DR} in a winter storm pattern can indicate a transition from rain to snow. By comparing differential reflectivity to reflectivity products, the forecaster can accurately pinpoint hail cores and updrafts cores.

8.5.2. CORRELATION COEFFICIENT (ρ_{hv}) (rho$_{DP}$) (CC), ratio. This measure examines all of the pulses within a radar bin to determine the distribution of power between horizontal and vertical planes. Values of 0.96 to 1 indicate low diversity, and is common in rain and dry snow. Values of 0.85 to 0.95 indicate a large diversity, consisting of different precipitation types such as hail and aggregates. Below 0.85 other phenomena are likely taking place, such as ground clutter, insects, birds, chaff, and at even lower values, tornado debris. It should be noted that correlation coefficient does not work well in weak reflectivity regimes, such as in very light stratiform rain or drizzle.

8.5.3. DIFFERENTIAL PHASE SHIFT (ϕ_{DP}) (phi$_{DP}$), deg. Measures the difference in phase between the horizontal and vertical plane

Figure 8-3. Spectrum width has been historically neglected for forecasting use, but one unique capability is the location of debris clouds, especially in ambiguous hook echo regions. Shown here is the debris cloud of the 10 May 2010 Little Axe tornado ("X") as seen on reflectivity (left) and spectrum width imagery. At this point the tornado was beginning to produce EF4 damage. Debris clouds tend to contain metal, which is highly reflective and results in the ball-shaped appearance on the reflectivity hook echo. Also, the lofting of debris with significantly different shapes, sizes, and terminal velocities causes the debris cloud to exhibit high spectrum width. The author's house was about 300 ft north of the EF4 centerline; he fled several miles south just minutes before the tornado's arrival.

at a particular radar bin. Radar energy may slow very slightly as it passes through water and ice, since the speed of light in those substances is different from that in air. While this slowing is not readily detectable, it can be immediately apparent when the waveforms of the horizontal and vertical phases are compared. If the waves are out of sync, that is, out of phase, this indicates that one of the phases was slowed by water, air, dust, or other particles more than the other phase.

The phase difference is additive with range, with the phase shift at a given bin prone to greater and greater variance with distance. This can be likened to building a wall by stacking bricks vertically, one atop the other without mortar; since the top of the wall is entirely dependent on the thickness of bricks underneath, the bricks get more and more uneven with increasing height. This same quality gives the differential phase shift display a radially streaked appearance with greater "noise" the further one goes from the radar. It makes the product difficult to use, so forecasters generally do not use differential phase shift product, relying instead on a different base product called specific differential phase, listed below.

8.5.4. SPECIFIC DIFFERENTIAL PHASE (K_{DP}), deg per km. This quantity examines the differential phase shift (see above) and simply shows the amount of change in phase shift at each bin along the radial, rather than the total shift. This is expressed as phase change per unit distance. As a result, the specific differential phase product helps highlight where phase shifts are occurring and where they are most intense. Positive K_{DP} indicates that energy has been slowed mostly in the horizontal plane due to hydrometeors dominated by horizontally oblate particles like rain. Negative K_{DP} indicates slowing of energy in the vertically plane, and suggests vertically oblate particles like conical ice.

It has been found that K_{DP} is a very effective indicator for high rainfall rate when very high base reflectivity values are correlated to high K_{DP}. In such a situation, it indicates significant backscatter from particles that are almost entirely raindrops.

8.5.4. DERIVED PRODUCTS. The base products can be used to construct a set of derived products. One such derived product is the Hydrometeor Classification Algorithm (HCA), which actually maps the information to a specific precipitation type, yielding a color-coded radar image of precipitation types. The dual-polarization radar system also offers a Melting Layer Detection Algorithm (MLDA) and a Quantitative Precipitation Estimate

(QPE) product yielding improved rainfall rates compared to older WSR-88D units.

8.6. Problems and pitfalls

In spite of the massive amounts of useful data that a radar provides to a forecaster, the images are sometimes deceptive or require special interpretation. Some of the potential problems are described here.

8.6.1. ELEVATION. One of the most common mistakes made by beginners is not understanding the conical nature of the radar beam. It does not scan perfectly horizontally, but is tilted upward slightly to avoid ground interference. The curvature of the earth also adds to the height of the beam as distance increases. Consequently, the radar sweeps the sky conically, and echoes close to the radar are at low elevations, and those far away are at high elevations. With a typical "lowest elevation" radar scan, the beam is already 10,000 ft above the ground by the time it is 90 nm from the radar. This means that low-level signatures like hook echoes, low-topped precipitation, and outflow boundaries will not be seen at distant ranges because the radar "overshoots" them.

8.6.2. BEAM SPREADING. The radar beam is not like a pencil-thin laser beam. On the WSR-88D it has a width of 0.96 deg. This means that at increasing distance, the radar is sampling larger and larger bins of the atmosphere, with the width increasing by one

Figure 8-4. Beam height caused by excessive range is a significant problem with radar meteorology. In this example, the same stratiform rain situation is viewed from two radar perspectives. On the left side, the Great Falls radar shows that the city of Great Falls is being affected by bands of rain. On the right side, the Glasgow radar miles away fails to detect any rain that is further than about 100 miles from the radar, giving the appearance that Great Falls is free of rain. *(2 April 2011 / 1620 UTC)*

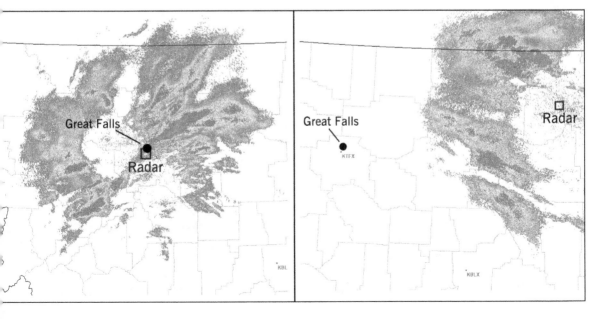

mile for every 60 nm of distance. Since reflectivity and velocity are all averaged within this bin, locally intense values of reflectivity or velocity are more likely to be lost. The larger beam also means the radar presentation is coarser at these distances, washing out fine-scale signatures.

8.6.3. RANGE FOLDING. The radar must send hundreds of pulses every second to build an image. There is a danger in doing so, because if the radar sends a new pulse while an echo is in the process of returning to the radar, the radar will interpret this echo as being caused by the new pulse. As a result, distant echoes will appear on imagery at a false position much closer to the radar. The radar operator can minimize this effect by selecting a lower pulse repetition frequency.

8.6.2. STORM MOTION. Consider a storm moving toward a radar unit at 35 kt, and that embedded in this storm is a mesocyclone which has a rotational velocity of 50 kt. This means that the forecaster analyzing the couplet will see a velocity of +15 kt on one side of the couplet and −85 kt on the other side. The forecaster can correct for this by using a storm relative velocity product. One instance where a storm relative velocity product is not appropriate, however, is when assessing ground-relative winds, such as when analyzing a bow echo to measure wind potential.

Figure 8-5. Velocity aliasing as seen on a derecho which affected southwest Missouri in 2009. The large swath to the west of the radar is a fast-moving MCS and the trailing stratiform precipitation area. The velocities are primarily inbound, reflecting fast movement toward the radar site, but shows a swath of high outbound velocities. By recognizing aliasing and using velocity de-aliasing algorithms, the forecaster can make sense of the pattern. *(8 May 2009 / 1230 UTC)*

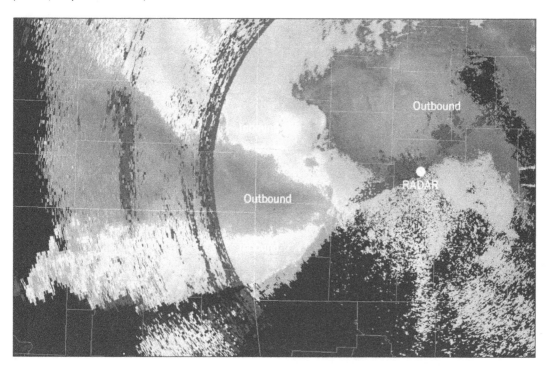

8.6.4. VELOCITY ALIASING. The WSR-88D radar measures velocity by analyzing the received energy as a train of waves, each wave forming one sampling "gate". Velocity is detected whenever a particular gate's waveform is out of phase with the waveform from the previous pulse. Since it takes two measurements of a single wave to accurately reconstruct it, the wave cannot shift by more than a half-wavelength during the interval or this will cause a computational error. Consequently the radar has a "speed limit" on atmospheric targets imposed by the PRF and wavelength. This speed limit is like a car speedometer calibrated from 0 at the bottom of the dial to 100 mph at the top of the dial, with the left side used for forward movement and the right side used for reverse movement. If the car goes forward at 110 mph, then the needle will rotate past 100 mph and show a -90 mph reverse reading.

The maximum unambiguous velocity of a radar is given by $V_{max} = PRF*\phi/4$, where ϕ is the wavelength in meters. For example, the WSR-88D's wavelength is 10.7 cm. If it is operating at a pulse repetition frequency of 1000 Hz, the maximum unambiguous velocity, or Nyquist velocity, is 27 m s^{-1} (53 kt). This yields a usable sampling range, also known as the Nyquist co-interval, of -53 to 53 kt. By increasing the PRF, the Nyquist velocity increases, giving more accurate velocity data, but the potential for range folding increases. This is known as the "Doppler dilemma".

Aliased velocity returns are extremely common in severe thunderstorms. Forecasters should be alert to whether they are using aliased or dealiased data, and look at surrounding velocity signatures and conceptual knowledge of the storm whether an aliased gate is being seen. Fortunately, computerized processing techniques known as dealiasing algorithms are widely used on radar display software to help locate and correct aliased parts of the image. These are often not available on web-based radar imagery, which often presents aliased images. If a dealiasing algorithm is not available, the aliased velocity values can be mentally "flipped" back taking into account the Nyquist velocity for that radar VCP, i.e. if an aliased velocity is 50 kt and the maximum velocity of the radar is 60 kt, then the true velocity is -70 kt.

8.7. Severe weather signatures

Systematic study of radar images over many decades has revealed a number of indicators of severe weather. An overview of these techniques requires an in-depth study of thunderstorms

NON-SEVERE   SEVERE

Figure 8-6. Reflectivity gradients. A non-severe storm tends to have low reflectivities, with the core well within the center of the storm. A severe storm often has high reflectivities, with a core shifted toward one side of the storm producing a strong reflectivity gradient on that side. It also shows reflectivity patterns that suggest a strong wind circulation (a hook, etc).

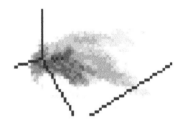

Figure 8-7. A tornadic storm in Texas, as seen on a reflectivity image. Note how the storm shows strong suggestions of a circulation, with a hook echo evident and a strong core near the south side of the storm.

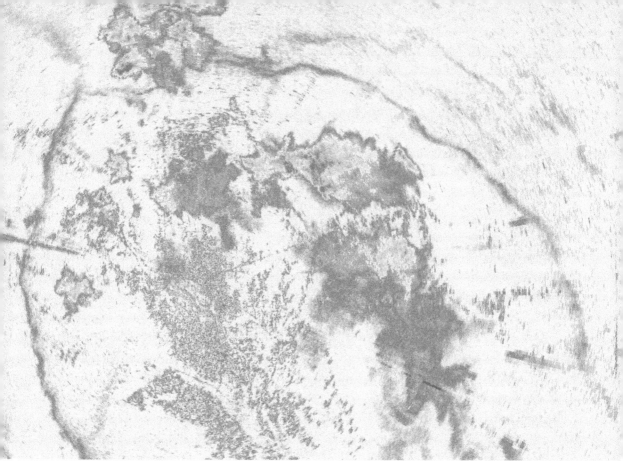

Figure 8-8. Dramatic example of outflow boundaries as seen in the Norman, Oklahoma area, using radar reflectivity from a specialized research radar. In addition to one large outflow bubble, there are also numerous old boundaries scattered across the area. On the majority of storm days, the question is not whether minor boundaries exist but how strong they are and whether radar, satellite, and surface data are able to detect them. In many cases they are not as prominent as seen here. Note that the highest reflectivities are mostly concentrated toward the centers of the echoes, indicating non-severe modes. *(26 May 2010 / 2236 UTC / OU-PRIME)*

and could fill an entire volume. Here we will focus on the most important indicators.

8.7.1. HIGH REFLECTIVITY. The obvious primary indicator is very high reflectivity values, which are indicative of heavy rain and hail. There are no threshold values that suggest severe weather, as even between two fairly identical storms reflectivities may be heavily influenced by rain density and ice processes rather than by circulations and hailstone size.

8.7.2. REFLECTIVITY GRADIENTS. The vast majority of weak thunderstorm cells contain a core that is positioned fairly well within the center of the echo as a whole. This gives it a concentric appearance. However, a core that has shifted strongly to one side (frequently the equatorward side) is a sign that strong circulations are developing within a storm.

8.7.3. HOOK ECHO. The well-recognized hook echo is an appendage caused by the wrapping of rain into a mesocyclone or tornadic circulation. It almost always consists of a cyclonic curve from base to tip. The hook echo is usually found on the right [NH] rear side of the storm, if it exists. Velocity products will show very high gate-to-gate shear values at the tip of the hook, and spectrum width products may show a small-scale maximum of spectrum width associated with the debris cloud.

8.7.4. BWER AND WER. The updraft region of a thunderstorm frequently develops a prominent echo-free area surrounded by precipitation material. In its weaker form this produces overhang of a strong echo over an area with weak echoes or none at all, and the area is known as a weak echo region (WER). If the signature is strong, the weak echo region can form a concavity that seems to punch into the bottom of surrounding reflectivity. This is a bounded weak echo region (BWER).

8.8. VAD/VWP wind data

At all times, NEXRAD radars collect data on winds throughout the entire scan volume. Even in clear air, dust, cirrus clouds, birds, and insects will scatter energy back to the radar, providing velocity information. This is mapped in a form known as the velocity azimuth display, if on a graph of velocity versus azimuth, or the velocity wind profile (VWP), if charted as velocity versus time. The preferred product is the velocity wind profile.

The radar processing system measures the average winds at each height and produces a product showing this wind profile. Although this data can sometimes be of questionable value, the availability every few minutes around the country can be of

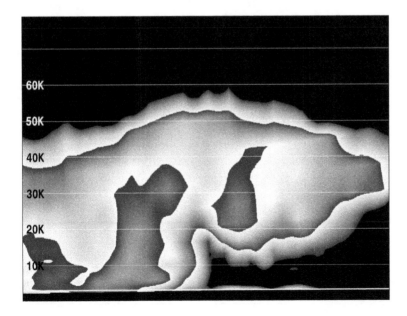

Figure 8-9. Supercell overhang on a storm near Piedras Negras, Mexico. This is a radar cross section looking north at the updraft area of the storm. The notch in the middle is a bounded weak echo region. *(24 April 2007 / 2334 UTC)*

significant value in keeping tabs on winds aloft when rawinsonde data isn't available.

Another type of radar technology similar to VAD data is profiler data. As of 2011 there were 35 profiler sites throughout the United States. Rather than being mounted with a dish antenna on a pedestal, a profiler looks directly upward, and can reach a height of 16 km. The data is readily available on the Internet at <profiler.noaa.gov>.

Chapter Eight
REVIEW QUESTIONS

1. Is conventional radar better at detecting cloud droplets or precipitation droplets?

2. Directly west of the radar is a couplet, with outbound velocity on the north side of the couplet and inbound on the south side. What kind of circulation is indicated?

3. Northeast of the radar is a couplet. The north side of this couplet has outbound velocity and the south side has inbound velocity. What kind of circulation is indicated?

4. Does a Doppler radar measure velocity along the radar beam or perpendicular to it?

5. Explain the basics of how a dual polarization radar works.

6. What kinds of degradation occur when trying to sample a storm at distant ranges?

7. What is an indicator that velocity aliasing has occurred?

8. How can a user of radar data correct an aliased velocity image?

9. What are some signatures on radar that might indicate a supercell?

10. Is the bounded weak echo region a manifestation of the updraft or the downdraft?

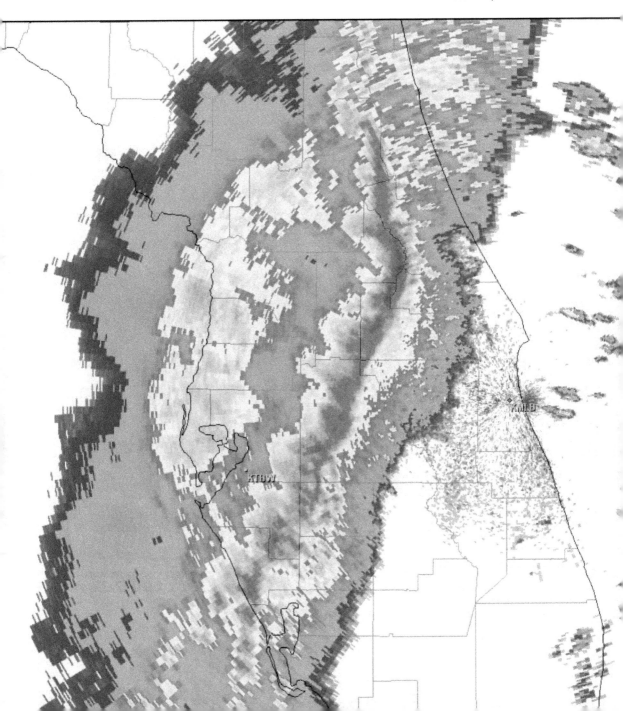

Figure 8-10. A mesoscale convective system, a squall line, moves across Florida. It caused extensive wind damage and was associated with the so-called Storm of the Century. *(13 March 1993 / 0601 UTC)*

9 CONVECTIVE WEATHER

When the weather becomes sultry and oppressive and giant towers billow into the sky, the forecaster must be familiar with convective forecasting principles. The foundation of this knowledge is the three key prerequisites for convection: moisture, a moist layer of sufficient depth in the lower or middle troposphere; instability, with a steep lapse rate within or above the moist layer to allow for a substantial positive lift area; and lift, with enough lifting of a parcel from the moist layer to allow it to reach its level of free convection (LFC). If all three of these conditions are met, the chance of convection is excellent. Greater amounts of moisture, instability, and lift will almost guarantee the formation of thunderstorm activity.

This is section is a basic overview of storm forecasting. For much more detailed coverage of storm structure, severe weather, and forecasting techniques, the reader is referred to *Severe Storm Forecasting* (2010), a title by the author which serves as a companion to this forecasting book.

9.1. Thunderstorm structure

A thunderstorm is made up of two components: the updraft and the downdraft. They are distinct features which are caused by different processes and are responsible for a variety of weather phenomena within the storm.

9.1.1. Updraft. The updraft is what fuels the storm. It consists of warm, moist air originating from storm inflow that is buoyant and rises rapidly. This frequently takes the appearance of a rapidly-building cumulonimbus tower. Exceptionally strong updrafts can take on rotation, sometimes producing a tornado under the base of the tower. As the tower grows, precipitation particles begin forming in the top of the tower, and eventually they fall out of the storm.

9.1.2. Downdraft. This feature is caused by precipitation particles accumulating at the top of the storm. The particles coalesce while others evaporate (chilling the air and causing it to sink); downward motion quickly begins occurring, forming the downdraft. It can fall directly back into the updraft, or in stronger storms, just downwind of the updraft (to the northeast). The downdraft, since it is sinking and causes cloud droplets to evaporate, appears visually as clear, bright air but may be filled (and darkened) with rain and hail. As the downdraft reaches the ground and spreads horizontally, it is known as outflow. The leading edge of the outflow is called the gust front. The

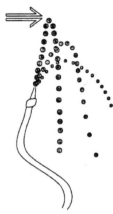

Figure 9-1. A thunderstorm updraft can be likened to the upward stream from a garden hose. The downdraft can be likened to the falling cascade of water droplets. If the upper-level winds are stronger than the low-level winds, the cascade (downdraft) will be carried downwind and will not "clog" the updraft. Therefore strong wind shear (difference in winds with height) supports long-lived thunderstorms. *(NOAA/NWS)*

Title image
Towering cumulus clouds rise over Oklahoma City. This is the last step before the formation of a cumulonimbus cloud. *(16 May 2010 / Tim Vasquez)*

Roman lightning protection

Among the Romans seal-skins were considered as an infallible preservative against lightning. Timorous persons anxiously crept into a tent covered with seal-skins when a storm was impending; and the Emperor Augustus, who seems to have been particularly afraid of thunder, always had a seal-skin near at hand. Arago relates that in the Cevenne Mountains, where Roman colonies were founded, the peasants carefully collect the skins of serpents, which they wear round their hats, and armed with this talisman, fearlessly brave the storm. Probably, these serpent-skins replaced among the people the more rare and costly seal-skins, which could only be purchased by the rich.

DR G. HARTWIG,
"The Aerial World," 1886

downdraft is responsible for the vast majority of the heavy rain, hail, and high winds in a typical storm. It is important to note that the downdraft has a strong tendency to choke out the updraft by falling directly onto it or undercutting it, which ends the life cycle of the storm. However, as we shall see, stronger storms circumvent this by handing off the updraft action to a different, newer updraft or by relying on strong upper-level winds to push the downdraft away from the updraft.

9.2. Multicellular storms

While there is a unicell storm, with a simple updraft-downdraft pair and considered the model for "popcorn storms", a true unicell is rare since the atmosphere tends to produce multiple updraft cells once the thunderstorm has matured. Even if a storm has one main updraft and one main downdraft, close inspection of what appears to be a singular blob of updraft or downdraft often shows "bubbly", discrete qualities. Therefore, in operational forecasting all storms are considered multicells, unless it appears to be a special variant such as the supercell or MCS.

9.2.1. MULTICELL CLUSTER. The most common type of storm is the multicell thunderstorm, which encompasses a broad spectrum. The cluster variety frequently appears as larger thunderstorms with many different updraft and downdraft cells in different stages of formation and decay and with a somewhat random distribution. The weaker multicell cluster usually forms the vast majority of summertime air-mass thunderstorms and tropical thunderstorms. A weak multicell cluster on radar plots usually looks like a

Figure 9-2. Cumulonimbus tower reaching skyward, forming the initial building block of the thunderstorm. *(13 May 2008 / 2245 UTC / Norman OK / Tim Vasquez)*

chaotic area of moderate-intensity cells, and cell competition is noticeably high.

9.2.2. MULTICELL LINE.

A boundary and a sheared environment favors a multicell line. These storms do not have random development along multiple gust fronts. Rather, activity concentrates along a very short line about 10-30 miles long, where the old cells die out on side away from the moist inflow, and newer cells sprout up closer to the moist inflow (usually the equatorward side). The older cells turn into rainy downdrafts and eventually dissipate, while the newer cells are composed of towering cumulus clouds. This line of towering cumulus comprises a flanking line. As the most active cells die and the flanking line towers mature, an apparent cycle of "growth down the flanking line" is seen. This process, called backbuilding, tends to propagate the storm eastward rather than northeast. The majority of springtime severe storms are multicell lines. They can produce weak tornadoes, very large hail, torrential rains, and strong winds.

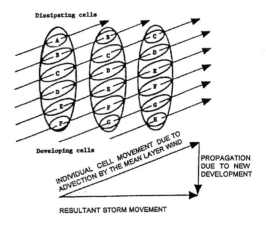

Figure 9-3. Concept of propagation of a multicell line. Even though the environmental winds may be flowing from southwest to northeast, new cell development ("backbuilding") occurs on the storm's flanking line to its south, with older cells to the north dying out. In effect, even though individual cells move toward the northeast, the storm as a whole moves eastward, to the right of the winds.

9.3. Supercells

The most dangerous type of thunderstorm in terms of the immediate threat to life and property is the supercell. It is defined as a convective thunderstorm with a clearly defined mesocyclone or mesoanticyclone. The storm is basically a

Figure 9-4. Multicell clusters in Louisiana on a summer day in 2008. About 40 years ago these would have been called "popcorn storms" or "air mass thunderstorms". Radar imagery here uses a special grayscale for maximum monochrome clarity. *(7 July 2008 / 2120 UTC)*

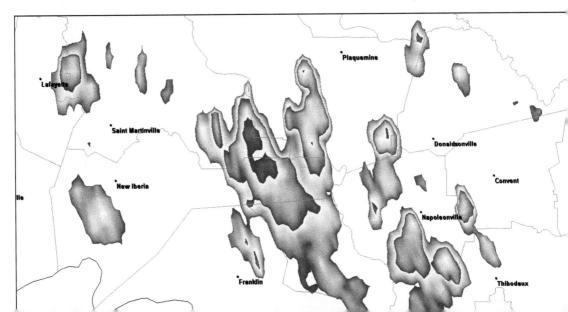

multicell which has exploited its environment to the fullest, shaping the surrounding air in a way which prolongs its own life and developing a single, long-lived updraft.

The term "supercell" was invented in 1962 by Prof. Keith A. Browning, who to this day is still an active researcher at Reading University in Great Britain. The supercell storm develops in areas of high instability with some degree of shear (and sometimes in low instability areas with substantial shear — thus the value of the EHI Index, to be discussed later). The initial life cycle is similar to that of the multicell line, however the flanking line towers show a tendency to grow very close together, even merging into one another, and the updraft base is fed by a strong, persistent inflow. Often within an hour or two, the storm will further organize and may produce lowerings and tornadoes, and on radar may show hook-like appendages on the southern side.

The classic (CL) supercell is what's usually seen in textbooks and diagrams. It involves a moderate amount of precipitation that falls out of the storm — just enough to help tornadoes spin up. These rain curtains get wrapped into the mesocyclone circulation, producing the hook echoes observed on radar. However, these rain curtains are eventually the tornado's demise. They get wrapped into the circulation and occlude the tornado. This type of storm results from high instability and large storm-relative shear.

When very little precipitation falls out of a supercell storm, the result is called a low-precipitation (LP) supercell. Since the rain curtains and rain shafts remain small in area, vast portions of the supercell are exposed for viewing. This makes it excellent for photography. However, while the lack of rain shafts means great photography, their absence also takes away the processes that help spin up a tornado. So while these storms may produce only small tornadoes, they can still produce huge hail. Low-precipitation supercells are inefficient. This can be attributed to dry-air entrainment with high instability, in which the additional dry air makes the storm even drier. Also, strong shear with weak instability can produce an LP supercell, in which most precipitation forms the anvil rather than falls to the ground.

The high-precipitation (HP) supercell is arguably one of the most violent types of storms on Earth, at least if the entire storm is considered rather than just the tornado. Copious amounts of precipitation and large hail often fall out of the storm. Weak shear is often what causes the storm to be a HP supercell, because it allows the precipitation to fall back into or just outside the updraft, resulting in rain-wrapped tornadoes. Where there is somewhat stronger shear, precipitation falls around the updraft,

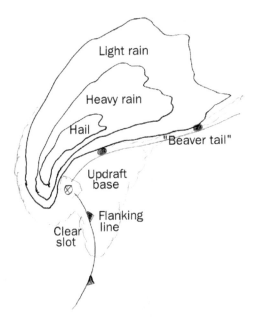

Figure 9-5. Structure of typical classic supercell. The wall cloud and potential tornado are located at the circle with the "X" in it. The small-scale fronts are indicated by the frontal symbology.

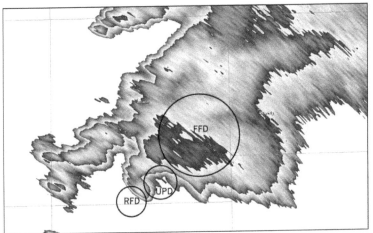

Figure 9-6. Parkersburg, Iowa tornadic storm in 2008. This shows the appearance of a supercell on radar, along with areas corresponding to the updraft (UPD), rear flank downdraft (RFD), and forward flank downdraft (FFD). The color palette has been mapped to a simple grayscale range as seen here for maximum monochrome clarity. Typical NEXRAD color schemes do not reproduce well in monochrome books. *(25 May 2008 / 2159 UTC)*

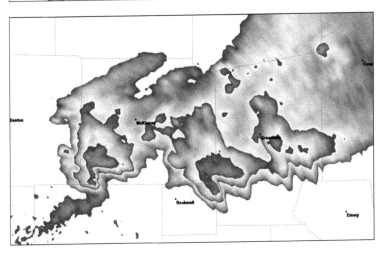

Figure 9-7. High precipitation supercells in North Texas. These storms occurred on the night of April 5, 2003, producing almost $1 billion in damage through the Dallas-Fort Worth metroplex due to large hail. Note the kidney-bean shape surrounding a "bears cage" feature. *(6 April 2003 / 0420 UTC)*

Figure 9-8. An LP supercell is essentially nothing but a chimney with little or no downdraft, as seen here. Many LP supercells are larger and more well-developed, however; this is either a young or miniature specimen. *(27 March 1997 / Tim Vasquez)*

but is wrapped into the mesocyclone, forming what is known as a "bear's cage" (the tornado is hidden inside the cage of rain).

9.4. Mesoscale convective systems

Thunderstorms may organize into systems that are larger than the scale of any individual storm cell. The processes within each cell and their interactions with one another shape the behavior of the entire system. Such a system of thunderstorm cells is known as a mesoscale convective system (MCS). Though the MCS is commonly considered to be a squall line, a complex of multicellular thunderstorms in Louisiana or a hurricane can be considered an MCS.

9.4.1. SQUALL LINE. A squall line is a long line of thunderstorms which extends for hundreds of miles in length. It is known for an exceptionally prominent two-dimensional structure. A classic squall line has its updraft on the east (front) side, with the downdraft on the west (back) side. This unique orientation and the linear nature of the gust front acting as a front-like wedge tends to push the updraft along rather than cutting it off, which means the updrafts are sustained for many hours. This results in a steady-state storm. Squall lines may rage for many hours or even days unless they move into an area of less moisture, unfavorable upper-level wind conditions, or higher stability.

Squall lines have a high amount of cell competition that is, there are so many cells along the line that it's difficult for any one cell to take over and become severe. This is why tornadoes are rare and short-lived, and hail (although common) remains rather small. However, the forward motion combined with favorable downdraft intensity can cause strong straight-line winds as the storm marches eastward.

9.4.2. Bow ECHO. Sometimes, a cell within the squall line accelerates ahead of the others. Such a bulge is called a bow echo, and on radar it has historically been referred to as an LEWP (line echo wave pattern). The bow echo is associated with a mid-tropospheric rear-inflow jet, entering the back side of the squall line eastward, and which then descends to the ground. This jet is driven by pressure perturbations in the squall line which go beyond the scope of this book. When the rear inflow jet reaches the ground, usually on a localized scale of tens of miles, the outflow increases and that part of the storm accelerates, with strong damaging winds in the outflow. A rear inflow notch may be seen on radar at this time. Enhanced vorticity where the left and right side of the rear-inflow jet reaches the ground produces

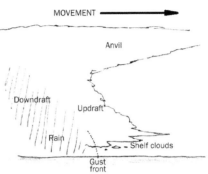

Figure 9-9. Two-dimensional slice through a typical squall line. The updraft tends to orient itself on the leading edge of the storm, with a trailing area of rain. Some small hail may occur along the leading edge of the rain. Straight-line hodographs with strong bulk shear tend to favor squall line scenarios.

Figure 9-10. Mesoscale convective system in Kentucky and Tennessee. This system was a squall line, and was also a derecho event. *(16 June 2009 / 1917 UTC)*

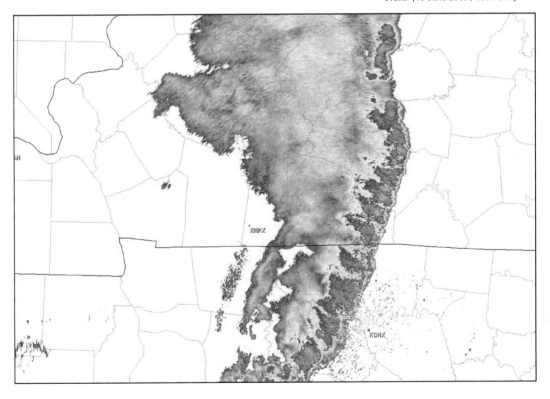

bookend vortices, which may spin up into tornadoes. Other vortices may form along the leading edge of the squall line, especially the poleward side of the bow echo, and may evolve into brief tornadoes.

9.4.3. DERECHO. Occasionally, a entire squall line that is fast moving, contains high outflow winds, or evolves into a large bow echo will produce a damaging swath of winds. This is known as a derecho. It does not refer to a specific storm structure but rather to a damaging wind event. Derechos can produce a swath of 100 mph winds 50 to 200 miles wide. In the United States, derechos are most common in the Midwest and Great Lakes states and are rare in tropical and subtropical latitudes.

9.4.4. MESOSCALE CONVECTIVE COMPLEX (MCC). The MCC is a special term given to mesoscale convective systems over the central U.S. (Maddox 1980) which have a specific amount of areal anvil coverage and coldness. For all practical purposes they are mesoscale convective systems and have no special status aside from large scale and intensity. The MCC was regarded as producing the majority of the warm-season precipitation in the Midwest, however due to the restrictive definitions and the need to consider

Figure 9-11. Bow echo in southwest Missouri on 0.5 deg radar reflectivity, showing a distinctive rear inflow notch (RIN) on the back side of the bow echo. This indicates the descending rear-inflow jet. Also noteworthy is a bookend vortex at X. *(8 May 2009 / 1226 UTC)*

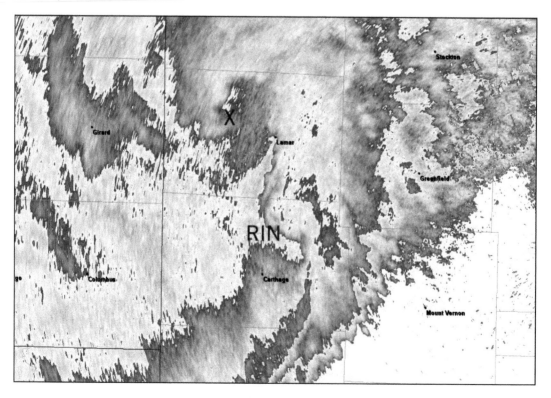

the entire spectrum of MCS systems, the use of the MCC term has diminished.

9.5. Wind profiles

The character of a thunderstorm is largely determined by its storm-relative wind profile. To evaluate storm-relative winds, we need the hodograph. The basis of the hodograph is quite simple: upper-level winds are always measured relative to the ground. When there is a 30 mph wind blowing from west to east across the plains, we assume it is blowing over barns at 30 mph, over treetops at 30 mph, and over lakes at 30 mph. However if there is a storm moving eastward at 40 mph, it would see this wind as moving from east to west (the opposite direction!) at 10 mph. Therefore one of the most important goals of working with a hodograph is to evaluate storm motion. Once storm motions have been established, the amount of shear, inflow, and ventilation can be determined. These help indicate what type of storm will occur.

Figure 9-12. Blank hodograph. This tool helps the forecaster perform important shear calculations. The scale is calibrated in units of wind direction (i.e. where the wind is originating from), but these are marked in a reverse orientation so that the wind vector (where it is blowing to) will appear correct. Shear calculations always consider where air is moving toward. To use the hodograph, the forecaster simply plots the wind direction and speed at a given station at all heights, and connects the resulting line in order of increasing height.

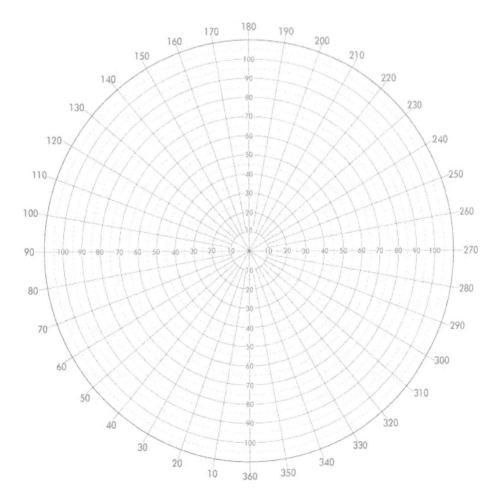

It is important to not just use the most convenient hodograph that's available. The hodograph must be modified to provide the best possible estimate of winds near the storm threat area at the time of initiation. Numerical model output, radar VAD/VWP winds, and profiler winds can all contribute greatly to this task.

9.5.1. STORM MOTION. In some cases, storm movement can be determined simply by studying radar animations. However the atmosphere rarely provides this luxury, requiring the forecaster to anticipate storm character before any activity even develops.

In general, the movement of a thunderstorm tends to represent an average of the winds at all levels through the troposphere. However it has been found that the lower levels of this depth are most important. Therefore the standard method of determining storm motion is to take an average of the 0-6 km winds. This can be done on the hodograph by visually averaging all the points between 0 to 6 km, ensuring that an even distribution of heights from 0 to 6 km are considered.

For typical storms, this techniques works quite well. Unfortunately storms often deviate from this movement and complicate the profile. They need to be planned for, and the deviant movement must be plotted on the hodograph.

For at least a decade, a rule of thumb has said that when normal storm movement is less than 30 knots, the deviant storm moves 20 degrees to the right and 85% the speed of the regular storm motion (thus the 20R85 rule). If normal storm movement is greater than 30 knots, the deviant storm moves 30 degrees to the right and 75% of the speed of the regular storm motion (thus the 30R75 rule). Unfortunately this rule of thumb sometimes performs poorly since it performs computations relative to ground speed.

A new method, called the ID method, calls for deviant motion to be evaluated not relative to the ground but to the 0-6 km shear vector of the storm. The forecaster simply draws the normal storm motion dot, draws a shear vector connecting the average 0-0.5 km and average 5.5-6 km wind vectors, and draws a line perpendicular to this line which intersects the normal storm motion dot. The deviant movement dot is plotted along this second line at a vector length of 7.5 m/s (15 kt) from the storm motion dot on each side of it. The deviant movement dot on the right side of the 0-6 km vector (facing the 6 km dot from the 0 km dot) is the right movement vector, and the other one is the left movement vector.

Still, storm movement is not an easy task and can be complicated by frontal boundaries, cold pools, topography, and

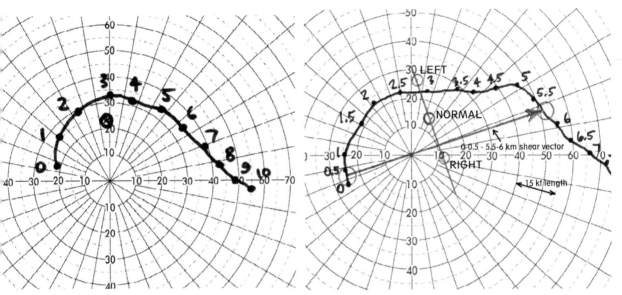

Figure 9-13a. Calculation of normal storm motion. The storm motion corresponds to the geometric center of the profile trace in the lowest 6 km of the atmosphere. This can be calculated mathematically, but it's faster to do it mentally by visualizing a spot that could "gravitationally" balance each of the points from 0 to 6 km. In terms of ground motion, this particular storm motion example illustrates cell motion moving north at 25 knots (originating from 180 deg and moving towards 360 deg).

Figure 9-13b. Calculation of deviant storm motion, ID method. A vector is drawn connecting the average 0-0.5 km and 5.5-6 km wind. Then a line is drawn perpendicular to this vector that intersects the normal storm motion. A right and left deviant movement vector is then plotted 15 kt on either side. It can be seen that right-moving storms will move eastward at 10 kt while left-movers will track northward at 25 kt.

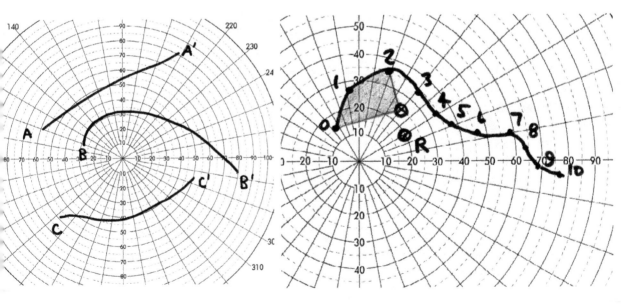

Figure 9-13c. Depictions of straight and curved hodographs. Trace A-A' is a straight hodograph that favors splitting storms and squall lines. Trace B-B' is a clockwise-curved hodograph that favors right deviant motion and contains positive storm-relative helicity. Trace C-C' is a counterclockwise-curved hodograph, rare on storm days, that favors left deviant motion and contains negative storm-relative helicity.

Figure 9-13d. Storm-relative helicity is the geometric area swept out between the storm motion vector and the hodograph trace between 0 and 2 km. The larger the area, the larger the storm-relative helicity. Note that if a storm begins moving along the right-movement deviant vector (labelled "R") the area swept out will be much greater.

unusual storm structures. Errors can also occur from poorly constructed hodographs using nonrepresentative winds, or unusual layers of winds in the atmosphere.

From here on, "storm motion" will refer to the movement of a storm along either the normal or the deviant movement vector. It's up to the forecaster to assess these possibilities and what they will contribute to storm development.

9.5.2. STORM-RELATIVE INFLOW.
By comparing the length of the vector between the storm motion vector and the lowest 2 km of the atmosphere, this yields the storm-relative inflow. The greater the inflow, the greater the feed of moisture into the storm. Storms with weak storm-relative inflow tend to be weak.

9.5.3. STORM-RELATIVE HELICITY.
The geometric area swept out between the storm motion vector and the lowest 2 km of the hodograph defines the storm-relative helicity. This is one of the key ingredients tied to rotating storms and tornado development. A computer can quantify these values, but a trained eyeball estimate can suffice quite well.

One configuration that can cause a larger area to be swept out is a curved hodograph trace, particularly one which is curved in the lowest levels of the atmosphere. This allows the hodograph trace to wrap more fully around the storm motion point. Backing winds at the surface and in the lowest level of the atmosphere can allow a relatively straight hodograph to take on curvature that increases storm-relative helicity values.

A straight-line hodograph tends to favor splitting storms, where a severe storm splits and one portion moves along the right deviant motion vector and the other moves along the left.

9.5.4. STORM VENTILATION.
By comparing the length of the vector between the storm motion vector and the upper part of the atmosphere, this indicates the speed of the anvil cloud and precipitation relative to the storm (i.e. the ventilation of the storm). Large storm-relative ventilation values are essential for a long-lived updraft.

9.5.8. STORM SPLIT.
Severe thunderstorm cells (particularly supercells) sometimes break in two, thought to be due to vertical updraft splitting by wake flow curl or precipitation loading. This is rare, but when it happens, the right split continues moving east and remains tornadic. The left split is weaker but can contain an anticyclonic tornado (very rarely though).

9.6. Tropical weather circulations

In the tropics, weather is overwhelmingly dominated by convective activity and a pronounced lack of baroclinicity. Occasionally a cold front will make it far south and produce genuine frontal weather, but such occurrences are rare. Furthermore, since pressure gradients are quite weak, forecasters must resort to streamline analysis to keep track of weather patterns. In spite of the paucity of highs, lows, and fronts, a forecaster can frequently uncover a smorgasbord of troughs, ridges, cyclones, anticyclones, outflow areas, and other disturbances. At the larger scale, there also a variety of weather regimes and circulations that dominate tropical weather.

9.6.1. EQUATORIAL TROUGH. A belt of strong solar heating in the tropics circling the globe defines the location of the equatorial trough. This is not a localized trough with a definite axis on synoptic weather maps, but is a generalized east-west area of low pressure throughout the tropics. In and near the equatorial trough there is a gradual upward flux of heat and moisture into the troposphere, and clouds and precipitation are widespread. This also gives it the name *tropical rainbelt*.

Figure 9-14. Equatorial trough in the Pacific Ocean. The general, large-scale convergence in the equator region describes the equatorial trough. (2 April 2011 / 1200 UTC / NOAA)

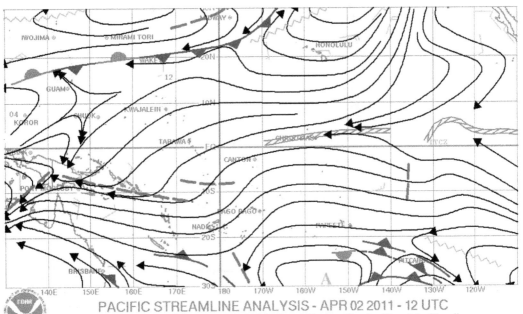

PACIFIC STREAMLINE ANALYSIS - APR 02 2011 - 12 UTC
KVM-70 U.S. Dept. of Commerce/NOAA/National Weather Service Honolulu, Hawaii

9.6.2. INTERTROPICAL CONVERGENCE ZONE (ITCZ). The ITCZ is by definition the zone where winds *converge* from north and south into the equatorial trough region. During the rise of the Norwegian cyclone theory, the ITCZ was considered to be an equatorial front which formed a continuous ring around the globe. However this convergence zone is severely disrupted over land areas by diabatic heating. As a result, the ITCZ is mostly an oceanic feature and is not necessarily co-located with the equatorial trough. When the ITCZ is away from the equator and is active it may be capable of producing waves, which move westerly at 10 to 15 kts. This is called an equatorial wave. When they occur, equatorial waves bring enhanced convection along the ITCZ. The ITCZ's poleward apex may also undergo breakdown, producing a tropical cyclone which may intensify if enough shear, warm ocean water, and Coriolis force are available.

Figure 9-15. Showers and thunderstorms along the equatorial trough. This photo was taken of Guam from the NASA Aqua/Terra satellite family. *(10 October 2010 / 1900 UTC / NASA)*

9.6.3. MONSOONAL TROUGH. The sun moves north or south with the seasons and the equatorial trough follows. In land areas, the equatorial trough is enhanced by and often overpowered by the strong heating of land surfaces. Warm-core barotropic lows develop in the region of strongest heating, focusing low-level convergence. As deeper and deeper moisture advects into these lows over a period of weeks, larger areas benefit from convective rains. The phenomenon in India is perhaps the most famous example of a monsoon, arriving in June and July on a typical year.

9.6.4. SUBTROPICAL RIDGE. The subtropical ridge is the descending portion of the Hadley cell (see Chapter 1). It comprises a belt of high pressure ringing the globe at about 30°N and 30°S. It is produced when air in the equatorial trough rises, moves poleward, converges due to the Coriolis force and a net accumulation of mass at this latitude, and sinks. The subtropical ridge's average position is 30 degrees latitude, and it contains vast amounts of subsident air. It is relatively dry compared to the air closer to the equator and skies are generally fair, though this may be offset by evaporation from oceans, evapotranspiration, moisture advection from other areas, and inversions which may cause clouds and haze to stratify.

Broad high pressure areas are established at the surface and these are considered warm-core barotropic systems. The poleward side of these subtropical highs are the primary source of tropical air in temperate latitudes.

Finally, subtropical ridges may sometimes be associated with the subtropical jet. If sufficient instability is present to produce storms, this may augment and prolong them due to the higher amounts of bulk shear.

Figure 9-16. Simplified diagram of the tropical circulation during the North Hemisphere summer. This shows the equatorial trough or "rain belt", along with the trade winds and subtropical highs. *(Tim Vasquez)*

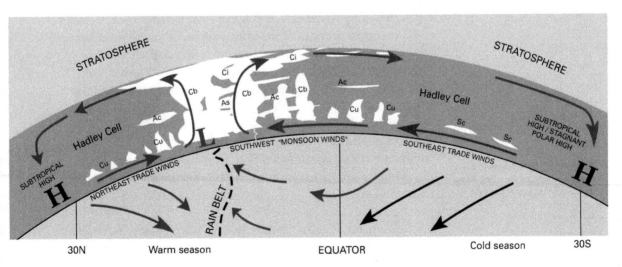

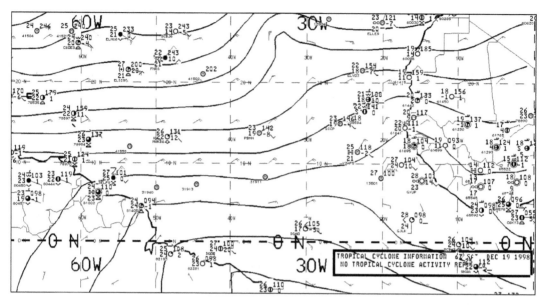

Figure 9-17. Trade winds in the Atlantic. These northeasterly winds were relied upon by early sailing ships to reach the Americas. This route went southwest to the Caribbean region, taking a little over a month. Ships took on supplies, then proceeded north up the Atlantic Coast. An old saying for the route out of England was "south 'til the butter melts, then west." *(19 December 1998 / 0600 UTC)*

Figure 9-18. Typical appearance of an easterly wave. This map uses wind plots and streamlines.

9.6.5. TRADE WINDS. The return circulation of air at the low levels from the subtropical ridge to the equatorial trough forms the trade winds. This wind dominates nearly all of the tropics. Since the equatorward-moving air deflects to the right because of the Coriolis force (left in the southern Hemisphere), the result is a large band of easterlies: northeasterlies north of the Equator and southeasterlies south of the Equator.

The areas of the trade wind belt with the deepest easterlies, usually within 15 deg of the Equator, are referred to as the deep easterlies. Wind is consistently easterly throughout most of the atmosphere. Further north and south are the "shallow easterlies", which are only 10 to 20 thousand feet deep. Though the trade winds exist at the surface, the air aloft tends to be west-to-east due to the proximity of the mid- and upper-level westerlies.

The coolest portions of the tropics are found in the subtropics over eastern ocean basins where the waters are relatively cold. This creates a cool-below-warm thermal setup in the lowest few kilometers of the troposphere, forming what is known as the trade wind inversion. The temperature can increase 10 deg C or more with height, and vertical cloud development is largely suppressed or favors extensive stratiform cloud layers. Areas within the trade winds with strong equatorward flow, particularly on the east side of subtropical high pressure areas, are associated with the strongest trade wind inversions.

9.6.6. EASTERLY WAVES. An easterly wave is a migratory wave-like disturbance in the tropical easterlies. It moves from east to west, having a cycle of about 3 to 4 days and a wavelength of about 1200 to 1500 miles. They often bring rain showers and in some instances may develop into tropical cyclones. The maximum intensity is usually in the lower and middle troposphere. Convective activity generally parallels the low level flow. The southern countries of western Africa are a major source of these waves — about 60 waves are generated there every year. These move westward into the Atlantic and may affect weather systems even as far west as the eastern Pacific Ocean.

Intense heating in the Sahara desert produces a semipermanent warm-core barotropic low. This is reflected aloft as a high pressure area. In turn, an easterly upper jet sets up south of this low. Minor baroclinicity (horizontal thermal instability) within this jet then causes easterly waves to develop. This wave moves westward with the trade winds.

Three different types of easterly waves have been identified, largely classified according to the slope of the trough with height. The stable wave is the most common type, which slopes toward the east with increasing height and moves slower than the prevailing flow. This produces enhanced low-level convergence and precipitation on its eastern side. On the other hand, the unstable wave slopes westward with height, and the convergence and precipitation field are found on the west side. A neutral wave stacks vertically and most of the precipitation is in the center of the wave.

9.6.7. Mid-Tropospheric Cyclone. These are cold-core barotropic lows which are caused either when a cutoff low becomes detached from the upper-level flow in middle latitudes or when an occluded low moves into tropical latitudes

Figure 9-19. Easterly wave near Iwo Jima in the Pacific, as seen on visible satellite imagery (center), the surface analysis (lower left), and the 500 mb chart (lower right). The box shape shows the extent of the visible image. The eastward slope with height suggests a stable wave. *(12 July 2010 / 0105 UTC)*

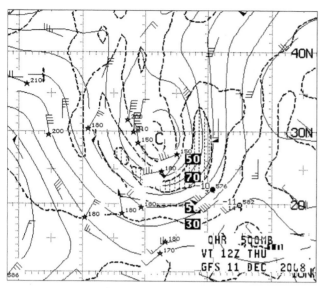

Figure 9-20. Kona low, a mid-tropospheric cyclone across Hawaii as seen on the 500 mb analysis. This storm produced 10 to 13 inches of rain on Oahu in just 12 hours. Mount Waialeale recorded a 4-day total of 24.70 inches. These storms occur primarily during the winter season. (11 December 2008 / 1200 UTC)

and becomes cut off from the mid-latitude westerlies. Most rain occurs on the eastern side of the cyclone, extending from 200 to 500 miles from the center, and the low may become stationary or drift erratically for days or even weeks. A good example of a mid-tropospheric cyclone is the "Kona" storm which is occasionally seen in the Hawaiian Islands in the winter.

9.6.8. TUTT. Occasionally the 300 mb or 200 mb analysis reveals a trough in the upper troposphere between subtropical ridges, particularly in the western parts of an ocean basin. The trough is often oriented SW-NE (Northern Hemisphere). These are referred to as TUTTs, or "tropical upper tropospheric troughs". A TUTT usually contains little significant weather. It may enhance the formation of surface disturbances, and in the Pacific it is considered as a source of tropical cyclone development. A TUTT may occasionally "close off", producing a TUTT low. These may move erratically or remain stationary. As the TUTT low increases in size, it often moves west-southwestward at about 10 kts. During tropical cyclone season, TUTT lows are important because they represent a source of upper-level shear, which can disrupt the balance of circulation within tropical cyclones.

9.7. Tropical cyclones

A tropical cyclone is the correct international term for a warm-core barotropic low which has intensified into a deep subsynoptic cyclone, largely due to the widespread release of latent heat. The American hurricane, the Asian typhoon, and the Austro-Indian cyclone all refers to the same thing: a tropical cyclone.

9.7.1. TROPICAL CYCLONE REQUIREMENTS. A tropical cyclone generally develops from a pre-existing disturbance like an easterly wave. It requires three key ingredients: a *warm moist surface*, an *unstable atmosphere, Coriolis force, and weak shear*. The surface is most important and should consist of ocean waters in excess of 80°F and 200 ft or more of water depth. The Coriolis force is also important in helping to prevent the system from filling. Tropical cyclones generally do not develop within 300 nm (5 deg) of the equator. Weak vertical shear, with deep layer shear of less than 20 kt (10 m s^{-1}) is required.

9.7.2. TROPICAL CYCLONE TYPE DESIGNATIONS. Tropical cyclones are categorized according to their *sustained wind speed*. Hurricanes require a sustained wind of 64 kt to be a hurricane or typhoon, and at least 34 kt to be a tropical storm. Once a storm reaches hurricane status, it can be further categorized according to the Saffir-Simpson scale, which provides for five different hurricane intensities according to the estimated sustained wind. The scales are best treated as a continuous spectrum of possible intensities, and other factors such as storm movement, tides, intensity changes, storm structure, and tornado activity can have a much larger impact on the resulting storm damage.

9.7.3. STRUCTURE. The tropical cyclone is defined largely by its extensive spiral bands. These are lines of convection that seem to flow with the inflow into the center of the storm. They contain heavy rain, thunderstorms, and strong winds.

Outer rain bands are lines of convection that form the outer spirals of the storm. The contain mostly heavy rain and thunderstorms.

Strong tropical cyclones often develop an eye, containing only clear air and some low clouds. It is not known exactly what causes the eye. One theory suggests that divergence aloft over the eyewall convection meets over the exact center of the storm, creating convergence and producing downward motion. It is also thought that the storm's rotation centrifuges mass away from the

The constancy of the trades

In the trade [wind belt] the constancy attains 80 percent. In no other regime on earth do the winds blow so steadily. Life has adjusted to this uniform wind stream in numerous ways. On many islands the towns lie on the leeward side, which affords better protection to shipping from wind and ocean. The author recalls from his days in Puerto Rico that there was always a great deal of excitement in the office during one of the rare "interruptions of the trade".

HERBERT RIEHL
"Tropical Meteorology", 1954

Figure 9-21. This tropical cyclone in Australia appears ominous at first glance on unenhanced infrared imagery (left), replete with spiral bands and outflow cirrus. The enhanced infrared imagery (right) shows the lack of a well-defined eye. The white spot in the center is not an eye, which should appear as a warm spot, but rather an enhanced area of cold cloud tops. This storm was Tropical Low 25U and had winds of 25 kt gusting to 45 kt. In the North American scheme, this makes it a tropical storm *(1 April 2011 / 2332 UTC)*

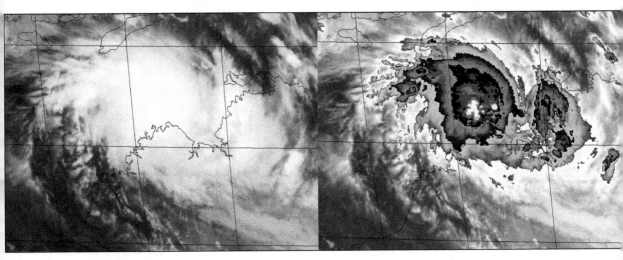

center, causing low-level divergence which produces downward motion.

Surrounding the eye is the eyewall. This is the innermost area of intense convection, often forming a tight ring. It contains the highest sustained wind speeds within the storm and the heaviest rain.

Very strong tropical cyclones may show a double eyewall, which is a concentric ring of convection within a larger eyewall. This occurs when the spiral bands concentrate and form an outer eyewall. The divergence aloft produced by the outer eyewall produces convergence aloft over the inner eyewall, which eventually causes it to dissipate. The outer eyewall then replaces the inner one. The tropical cyclone often weakens for a short time as this happens.

On satellite imagery, the tropical cyclone is defined by a large area of outflow cirrus. This is an area of high clouds, most often seen on infrared satellite imagery, that looks like a plume of cirrus expanding away from the storm. It often makes the storm appear much larger than it really is. Outflow cirrus is produced by advection of moisture outward by upper-level divergence created within the storm as a whole.

Figure 9-22. Eyewall of Hurricane Igor, the strongest hurricane of the 2010 season. At this time it was a Category 4 storm with winds approaching 155 mph. *(NASA/ GSFC)*

9.7.4. TROPICAL CYCLONE CHARACTERISTICS. The worst winds in a tropical cyclone are usually confined within the eyewall. The tropical cyclone contains enormous amounts of precipitation,

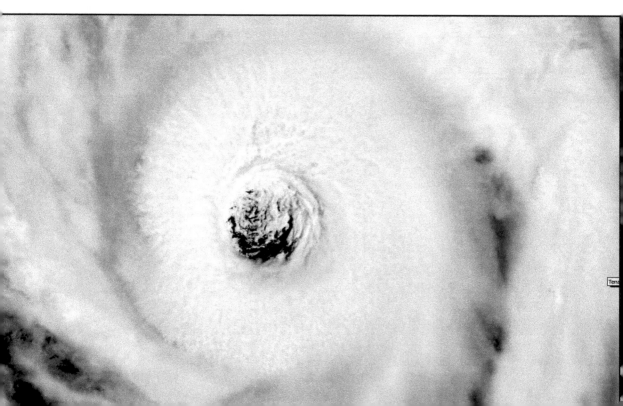

which may pose a serious flooding risk if the storm becomes stationary.

Tornadoes occur in many hurricanes. Landfall is usually the catalyst for tornado production; as friction begins causing winds to diminish, these weaker winds coupled with the stronger winds a few thousand feet aloft creates strong low-level shear that favors tornado development. Since instability is fairly weak and low in altitude compared to typical tornadic storms, hurricane-spawned supercells usually are small, low-topped, and tend to occur during the daytime with the best insolation. This, coupled with the fact that they are rain-wrapped and move quickly makes detection and warning very difficult. A climatological study of hurricane-produced tornadoes found that they usually occurred on the outermost bands (50 to 200 miles from the center) ahead of and to the right of the storm's movement.

In September 1967 Hurricane Beulah produced 141 tornadoes in southeast Texas, most of them small and short-lived. Hurricane Gilbert in September 1988 brought several large tornadoes in south Texas which damaged parts of western San Antonio and were videotaped by storm chasers near Del Rio. In August 1992 Hurricane Andrew produced 62 tornadoes.

9.7.5. MOVEMENT. The tropical cyclone has a tendency to move within the deep-layer flow in which it is embedded. Since this is usually within the trade wind belt, this constitutes a westward drift. However, it has long been known that tropical cyclones gravitate towards any synoptic-scale troughs at higher latitudes, while ridges tend to keep the storm equatorward.

The tropical cyclone is also affected by what is known as the beta effect. Absolute vorticity is a conserved property, meaning that if a parcel moves southward [northward], its relative vorticity must increase [decrease] to compensate for the decreased [increased] contribution of positive rotation from the Earth. In the northern hemisphere, a tropical cyclone spins cyclonically, which means the southward [northward] moving current is on the west [east] side. As a result, the storm will contain a maximum of positive relative vorticity on its west flank and a minimum on its east flank. The rotation from these vorticity centers, known as beta gyres, couples to produce a poleward "current" of air. Combined with the pre-existing westward drift, this produces a northwestward propagation of the tropical cyclone.

Storms are also affected by the release of their own latent heat, and may propagate towards where the strongest latent heat release is occurring.

Hurricane terminology

Invest. A weather system, which may not yet have tropical cyclone characteristics, which is subject to collection of additional data.

Tropical disturbance. In this stage, sustained winds are less than 34 kts (39 mph, 17 m s⁻¹), with an open circulation (no closed isobars).

Tropical depression. Sustained winds are less than 34 kts (39 mph, 17 m s⁻¹), but there is a closed circulation (closed isobars). The storm is assigned a number.

Tropical storm. Sustained winds are at least 34 kts (39 mph, 17 m s⁻¹) but less than 64 kts (74 mph, 33 m s⁻¹), with a definite closed circulation. The storm is assigned a name.

Hurricane. Sustained winds are at least 64 kts (74 mph, 33 m s⁻¹).

Supertyphoon. This term is used by the U.S. Joint Typhoon Warning Center in Guam to designate a typhoon with 1-minute sustained winds of 130 kts (150 mph, 65 m s⁻¹) or greater.

Storm tracks may show small-scale wobbles, known as trochoidal motions. These are mesoscale eccentricities in the track that may be misinterpreted as a change in direction of the storm. They are believed to be an effect of embedded mesoscale vortexes which circulate around the storm. If they are detected, meteorologists try to eliminate them when fixing the storm position. This may give the appearance of official forecasts not reflecting a change in the storm's movement, when in fact it is believed that a trochoidal motion is at play.

Finally storm tracks may show great wobbles due to the presence of another tropical cyclone nearby. This is called the Fujiwhara effect, in which the two storms have a tendency to rotate around one another.

9.8. Tropical cyclone forecasting

Extrapolation has long been a favored technique of tropical cyclone forecasters: grab a hurricane plotting chart, update the position of the storm, and attempt to project the movement forward. The rapid advances in numerical modelling have been slow to take hold with hurricane forecasting because of the limited

Figure 9-23. Enhanced infrared satellite image of Hurricane Ivan in 2004. The brightest areas near the eye of the storm correspond to very cold cloud tops, revealing the convective bands near the storm's center. Much of the areal extent of the enhanced shading is actually dense outflow cirrus near the tropopause and does not represent showery weather or even the extent of strong winds. This is a common mistake made by the general public when trying to interpret these images and gives the impression that the damaging swath of the storm is much larger than it really is. *(11 September 2004 / 0231 UTC)*

ability to sample the storm and the sheer complexity of processes within.

9.8.1. STATISTICAL MODELS.

Owing to their simplicity, the very first models were statistical. The National Hurricane Center developed the HURRAN model in 1969, which simply used historical analogs to find the most likely path of the storm. This was improved a few years with CLIPER, which continues to be used to this day, albeit mainly to provide a baseline for other forecasts. It takes into account not only the growing body of climatological storm tracks but also persistence. A similar model which forecasts intensity instead of position is SHIFOR.

A number of hybrid dynamical-statistical models were developed, starting with NHC73 for forecasting track.

9.8.2. SIMPLE DYNAMICAL MODELS.

Simple dynamical models for hurricanes were fielded as early as the 1970s. They took into account the field of winds, temperature, and moisture around a storm but did not actually solve equations of motion. Since they have statistical elements they are often referred to as hybrid models. The earliest examples are the QLM and MFM models.

The first mature hybrid model was the BAM (Beta Advection Model), introduced in 1987. It used layer-averaged winds within the global GFS/GSM model to steer the storm, and adds a correction for the beta effect.

A few years later, the LBAR (Limited Area Sine Transform Barotropic Model was introduced), which was similar to the BAM but took into account assumptions about the storm vortex and solved barotropic equations for motion.

9.8.3. FULL DYNAMICAL MODELS.

Full dynamical models attempt to model all elements of wind, moisture, and temperature throughout the tropical cyclone. Tropical cyclones were ingested into most global forecast models in the 1980s, though these models were woefully incapable of resolving any structure within the storm.

Due to the small scale of the tropical cyclone, nearly all current dynamical models have to be fed a synthetic, or bogussed vortex that best represents the structure of the storm. The basic way of doing this is to manually create a grid of winds representing the storm and feed it into the model. However newer models such as the GFS are capable of allowing technicians to simply adjust the vortex to the new location, a scheme known

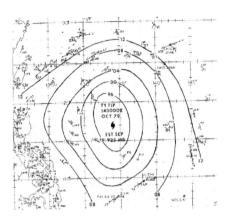

Figure 9-24. Supertyphoon Tip, the deepest typhoon in recorded history, showed a large pressure field that affected much of the western Pacific region. These isobars were drawn using ship reports only.

First hurricane encounter

The late 15th century brought the first hurricane encounter to European explorers. In 1494 Christopher Columbus made his second trip to the New World to establish a second Spanish settlement. It was named La Isabela and was founded on the northern coast of the Dominican Republic. Unfortunately within a couple of years two hurricanes hit and sank at least eight ships. In 1501 Peter Martyr wrote: "This strong south by southwest wind reached the city and the three ships that were alone at anchor. Without any perturbation of the water or surge of the sea, it broke their cables, gave them three or four twirls, and submerged them on the bottom." The losses coupled with diseases, parasites, and exposed anchorage led to the complete abandonment of La Isabela within four years.

Steering a hurricane
If no modelling information is available on tropical cyclone movement, the storm can be moved in the direction of the 700 mb flow, or for developed hurricanes and typhoons, the 500 mb flow.

as vortex relocation. Some models such as the GFDL are designed to spin up a vortex at a designated location before the model run begins. Finally, one model, the ECMWF, because of its high resolution is able to treat tropical cyclones nonsynthetically its model runs.

Mesoscale dynamical models are now available which can simulate the eyewall and convective bands. One of the best of these mesoscale models is the HWRF (Hurricane WRF Model), an implementation of the WRF described in the Prognosis chapter. It was fielded in 2007. The potential for this model to ingest radar observations in the near future brings enormous potential for accurate hurricane forecasts close to landfall.

Though full dynamical models offer a powerful tool for forecasting storm movement, meteorologists need to not just fixate on comparing storm tracks. The actual wind fields depicted by each model can be compared with actual observational data to help determine which solution is right.

Figure 9-25.   SLOSH is a numerical storm surge model used in emergency preparedness operations for the landfall of hurricanes. Using data about a hurricane and topography it attempts to predict the inundation of water along coastlines. *(NOAA/NHC)*

9.8.4. ENSEMBLES. Experimentation is also ongoing with ensemble forecasting, in which multiple model runs are made with slightly different initialization conditions. Multi-model ensembles can also be constructed from entirely different models.

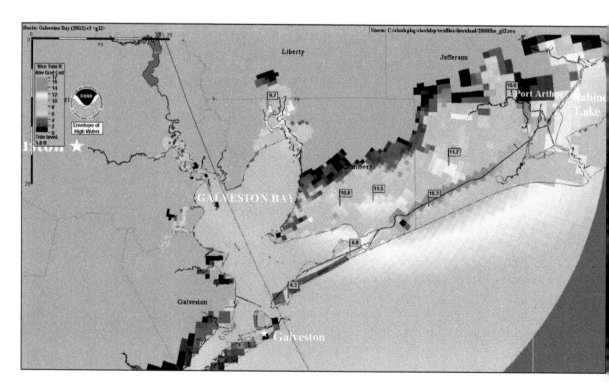

A consensus solution averages the ensemble models together to find the best fit for storm motion.

Overall, consensus ensembles provide the most accurate tool for forecasting storm motion. It is still up to the forecaster to check the ensemble members for errors and initialization problems.

Chapter Nine
REVIEW QUESTIONS

1. What are the building blocks of a thunderstorm, and what are they composed of in terms of clouds or precipitation?

2. Since the LP (low-precipitation) supercell rarely produces tornadoes, is it dangerous? If so, why?

3. Where is the updraft found relative to the squall line?

4. Is damaging weather more likely in a squall line if it organizes and solidifies or if it breaks up into isolated cells?

5. Why do strong winds aloft favor severe weather?

6. Can tropical weather best be described as baroclinic or barotropic?

7. Where is an easterly wave found, in relation to the equatorial trough, the trade winds, and the subtropical high?

8. You arrive in Hawaii for a vacation in February, and you find that heavy rains have been plaguing the islands for the past week. What tropical weather feature might be responsible?

9. Name at least three ingredients that are required for tropical cyclone formation.

10. Looking at satellite imagery, what feature gives the impression that a tropical cyclone is larger than it actually is?

Flying in the eye

Here was one of Nature's most spectacular displays. The eye was a vast coliseum of clouds, 40 miles in diameter, whose walls rose like galleries in a great opera house to a height of approximately 35,000 feet where the upper rim of the clouds was smoothly rounded off against a background of deep blue sky. The sea surface was obscured by a stratocumulus undercast except for two circular openings on the east and west sides of the eye, respectively. Clouds in the undercast layer were grouped in bands which spiralled cyclonically about each of these openings or clear spots, both of which were approximately five miles wide. This horizontal alignment of clouds suggested the possibility that two separate small eddy circulations were present within the eye envelope. In the geometric center of the eye the stratocumulus undercast bulged upward in a domelike fashion to a height of 8,000 feet. Light turbulence in the tops of this dome was comparable to that in ordinary ocean cumulus. The walls on the eye to the west side were steep, either vertical or overhanging, and had a soft stratiform appearance. On the east side however clouds were more of a cumuliform type with a hard cauliflowery appearance.

R.H. SIMPSON, 1952

10 PROGNOSIS

Without a thorough understanding of what is occurring in a forecast area, a forecaster is doing little more than divining the future, or at best, is simply regurgitating a forecast tool. This connection with the actual weather, and its comprehension and understanding, is known as *diagnosis*. Ironically, this technique has become a significantly neglected aspect of operational meteorological science. This is largely a result of the strong emphasis on numerical modelling, parameterization, and analysis in meteorology journals. On a day-to-day basis, it's not surprising to encounter a forecaster who painstakingly isopleths the surface map, reciting the importance of analysis, and then dives straight into the numerical models and gridded fields without reviewing other charts and tools.

As a result, the forecaster may have an intimate understanding of *what, when, where,* and *how* of a weather system and its evolution, but questions of *why* yield unconvincing answers. This phenomenon was highlighted as far back in the 1977 with meteorologist Len Snellman's warning of a growing "meteorological cancer". A forecast that is heavily grounded in actual, observed data *and* a keen understanding of how all the processes are tied together is the foundation of a forecast.

10.1. The forecast process

The forecast process is made up of four key elements: observation, analysis, diagnosis, and prognosis. Observation is the sampling of the atmosphere's current state. Analysis is the process of putting all of these observation pieces together into a unified picture, like assembling a jigsaw puzzle. There may be many analyses that are part of the forecast process, including the surface chart, radar data, and profiler data. Diagnosis is the interpretation of that picture, synthesizing all analyses into one coherent picture of the atmosphere. Prognosis, in turn, is the technique of determining how that picture will change with time.

10.1.1. OBSERVATION. The observation process is straightforward and is discussed in detail in Chapter 2. This is the collection and dissemination of atmospheric data samples, including surface reports, radiosonde observations, and profiler samples. Nearly all of this work is already done for the forecaster.

10.1.2. ANALYSIS. Analysis comprises the forecaster's first exposure to the conditions in the forecast area. The analysis is the process of studying and absorbing all of the details shown by the chart.

Title image
Forecasters at the Storm Prediction Center in Norman, Oklahoma in 2008 look at upper level charts as they contemplate issuing a tornado watch for Arkansas. *(Tim Vasquez)*

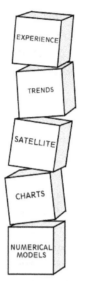

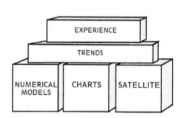

Figure 10-1a. Unbalanced forecast. Poorly skilled forecasters usually build from the bottom upward as shown here, forcing the next higher level to "fit" on top. Improper analysis of one product topples the entire forecast.

Figure 10-1b. A balanced forecast. Each forecast tool is used on its own merits then is synthesized with the other tools. Discrepancies and problems don't affect other tools.

The methodical procedure of hand analysis is a very important part of analysis, as it helps the forecaster extract patterns from the data and by its very design it prevents areas of the chart and small-scale features from being skipped. Not all fields on each chart require hand analysis, since this can lead to a very laborious and time-consuming analysis process, but the forecaster should select one or two important fields relevant to the forecast on each chart and make time to hand-analyze them. The remainder of the isopleths can be computer-drawn as needed.

A troubling development in recent years is the use of 00-hour model fields, gridded fields, or reanalysis charts for the "current weather" analysis. This is bad form, particularly for mesoscale analysis, because the forecaster is viewing a heavily processed form of observed data. In fact, such fields are often blended with forecast panels from previous model runs in order to provide properly balanced fields suitable for execution of a model run. This means that important details may be smoothed out. If synoptic-scale forecast work is being done, it may be acceptable to use a combination of data plots with machine-contoured output, such as the DIFAX upper air charts distributed by NOAA/NCEP.

A wise forecaster puts all numerical models and gridded data aside initially and focuses on analysis tools: surface and upper air plots, satellite images, radar data, and much more. Even the simple process of looking out the window provides a valuable scrap of analysis information. The forecaster's goal must be to sample the current state of the atmosphere in as many ways as possible in order to understand patterns and processes across the forecast area, even if they aren't immediately understood. The forecaster also tries to evaluate the trends of these patterns and processes.

10.1.3. DIAGNOSIS. The process of diagnosis is that of mentally blending all of the analyses into a unified picture of the current state of the atmosphere. At this point, the forecaster has a three-dimensional visualization of the atmosphere. This should also integrate an understanding of how the atmosphere has changed over the past few hours or days, which yields a four-dimensional understanding. When the diagnosis is complete, the forecaster can explain what type of weather regime any given location is under, what type of stability, air masses, and clouds exist there, and what its relationship is to nearby fronts and jets.

Diagnosis is not a separate process in itself; rather it grows as the forecaster completes all of the analysis work. If a diagnosis is not yet achieved, then the forecaster must obtain additional sources of data to analyze. If this is not available, the

Why is forecasting so difficult?
Consider a rotating spherical envelope of a mixture of gases, occasionally murky and always somewhat viscous. Place it around an astronomical object nearly 8000 miles in diameter. Tilt the whole system back and forth with respect to its source of heat and light. Freeze it at the poles of its axis of rotation and intensely heat it in the middle. Cover most of the surface of the sphere with a liquid that continually feeds moisture into the atmosphere. Subject the whole to tidal forces induced by the sun and a captive satellite. Then try to predict the conditions of one small portion of that atmosphere for a period of one to several days in advance.

Author unknown

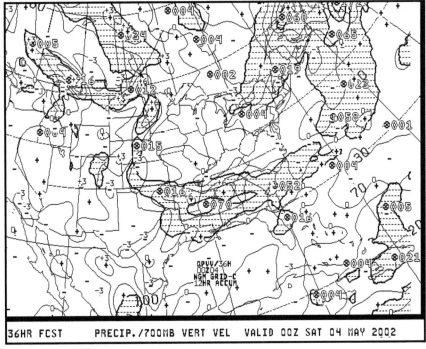

Figure 10-2. Vertical velocity. One way to look for vertical motion is to obtain it from numerical models. The 700 mb vertical velocity is available for all model runs and is a popular product. It solves the quasigeostrophic equation for omega (vertical motion). However, these fields are very noisy and can be difficult to interpret.

36HR FCST   PRECIP./700MB VERT VEL   VALID 00Z SAT 04 MAY 2002

forecaster must draw upon experience and expertise and develop a conceptual model of what is occurring in order to reach an accurate visualization of the processes across a forecast area. Finally, if a process cannot be explained at the synoptic scale, a mesoscale process is probably at work and further diagnosis should take place at that scale.

Figure 10-3. Model output can be blended with analysis charts as shown here on May 8, 2002, depicting the 1400 UTC analysis with projections of forecast frontal positions interpreted from the Eta and RUC runs. Seeing the analysis and forecast data on one page is a starting point for identifying discrepancies and trying to solve their underlying cause. *(Tim Vasquez)*

10.1.4. Prognosis. Though this is a forecasting handbook, the methods for accurately forecasting weather without numerical models are in fact quite limited. To properly forecast the weather, the physical equations that govern heat, motion, and moisture have to be solved, a process known as dynamic modelling. The mathematician L. F. Richardson made the first actual attempt at solving a prognosis this way, and it took two years of hand calculations to carry a single case study forward six hours.

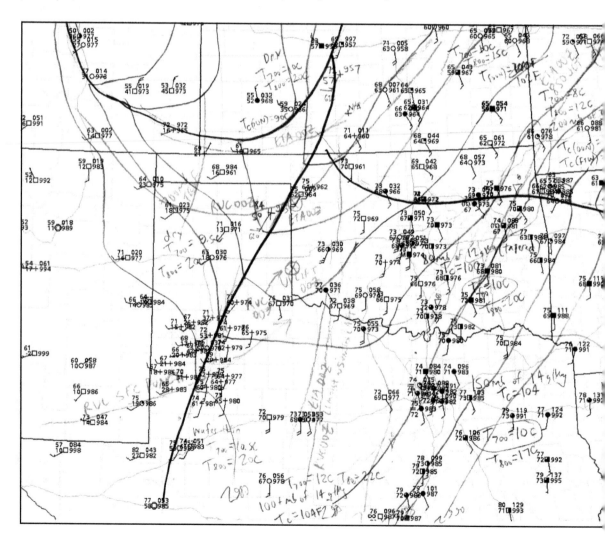

This leaves the human forecaster with only a repertoire of subjective forecast techniques, all of which are based on simplifications and generalizations of processes occurring in the atmosphere. These techniques include continuity, extrapolation, pattern techniques, decision trees, flowcharts, and rules of thumb (empirical rules). Though these were widely used until the 1960s, they are largely obsolete now for synoptic-scale forecasting given the dependable and accurate nature of numerical modelling. At the mesoscale, where models are just now beginning to show skill, subjective techniques do help in a supplementary capacity.

One area where humans excel is in forecasting with the thermodynamic diagram. If the patterns are not too complex and the forecast is limited to several hours in the future, a considerable amount of accurate prognosis can be done using a skew T log p diagram. This is because parcel motions are assumed to be primarily vertical, they adhere sharply to a number of key assumptions, and the diagram readily indicates the outcome in terms of weather phenomena.

Numerical models have matured significantly since their initial appearance, evolving into highly sophisticated forecasting systems. By their very design, numerical models usually excel at the broad scale patterns, but fail to resolve processes at small scales of time and space. Consequently, a major part of the forecaster's job is applying conceptual models of mesoscale processes to the picture given by the model output. For example, a numerical model does not indicate where the cold conveyor belt is in a baroclinic system, but a forecaster can interpret the carts, find the feature, and compare the model output with radar data to see whether it might have a greater influence on precipitation than expected.

Another area where humans excel, besides the small scale jobs, is with events occurring right now. Clearly it would be foolish to use models to predict what the weather will be in one minute; one only has to look at the surface observations. To forecast the weather twenty minutes from now, a forecaster might look at surface observations, then check the radar and satellite imagery. And to forecast weather two hours from now, a forecaster might look at all the observed data and blend it with their knowledge of mesoscale meteorology. This sort of short-range forecasting is a technique known as *nowcasting*. During the first few hours, it usually proves to be significantly more accurate than numerical models.

One other shortfall of numerical models is that they do not provide *context* readily obtained from a thorough analysis

Before chaos was recognized

I remember a talk that John von Neumann [great mathematician] gave at Princeton around 1950, describing the glorious future which he then saw for his computers. Meteorology was the big thing on his horizon. He said, as soon as we have good computers, we shall be able to divide the phenomena of meteorology cleanly into two categories, the stable and the unstable. The unstable phenomena are those which are upset by small disturbances, the stable phenomena are those which are resilient to small disturbances. He said, as soon as we have some large computers working, the problems of meteorology will be solved. All process that are stable we shall predict. All processes that are unstable we shall control. He imagined that we needed only to identify the points in space and time at which unstable processes originated, and then a few airplanes carrying smoke generators could fly to those points and introduce the appropriate small disturbances to make the unstable processes flip into the desired directions. The dream was based on a fundamental misunderstanding of fluid motions [which] fall into a mode of behavior known as chaotic. A chaotic motion is generally neither predictable nor controllable. It is unpredictable because a small disturbance will produce exponentially growing perturbation of motion. It is uncontrollable because small disturbances lead only to other chaotic motions and not to any stable and predictable alternative.

FREEMAN DYSON
Infinite In All Directions, 1988

Figure 10-4. Every good forecast office needs a good large format printer like the one seen here. Hand analysis on letter-sized paper can is tedious and strains the eyes. Due to the dot.com crash and the recession, there is a glut of large format printers at used equipment venues such as eBay. This Epson 1520 inkjet printer was used by the author from 1997 to 2009 and cost $500 new. It is capable of using 15-inch wide tractor feed paper, as seen here, which means that very large maps can be printed in landscape mode. This printer was replaced with a used HP Laserjet 5000N which cost $300. It is a laser printer which outputs onto 11 x 17" sheets. *(Tim Vasquez)*

and diagnosis. For example a model run might opt not to draw any precipitation in a given area. A forecaster who has diagnosed the soundings and low-level moisture fields will recognize the potential for precipitation to develop and will understand what factors might lead to development and why the models might be taking a pessimistic view. This kind of insight leads to a much more accurate, refined forecast.

10.2. Numerical model concepts

Although the equations for motion were worked out in the early 20th century, forming the basis for mathematical prediction of the weather, computer power was simply not available at the time. Researchers envisioned a large forecast center filled with mathematicians who worked as a team, subdividing parts of the atmospheric puzzle, making calculations with slide rules and paper, and reassembling the results to make a forecast. Though this idea never came to fruition, the development of the first operational computers in the late 1940s allowed this idea to be implemented as a completely automated task. Numerical weather prediction schemes were rapidly developed, and in 1958, the very first numerical model forecasts were disseminated to end users.

10.2.1. DYNAMICAL VS. STATISTICAL MODELS. Dynamical models simulate the atmosphere physically, modelling changes in air flow, heat, and humidity. They rely on the so-called "primitive equations" which govern motion in the atmosphere, as well as physical parameterizations which account for things that are too small for the model to simulate, such as thunderstorms and surface heating. On the other hand, statistical models relate the properties of the system to one another without relying upon physical equations. One type of statistical model is an *analog*, where a historic situation with similar qualities is used to provide a prognosis. Some models are both dynamical and statistical: MOS output, which is based upon dynamic models but then makes statistical forecasts from the output, is an example of this. Many hurricane forecast models are also statistical or hybrid models, since very little observational data is available within the storm and the finer details of convection cannot be adequately modelled.

10.2.2. MODEL DOMAIN. A dynamical model uses either a global domain or a limited domain. The global domain encompasses the entire world. The limited-domain model only considers a particular region of the Earth, such as North America only. The problem with the limited-domain domains is weather systems outside the domain cannot be forecast, and as a result, forecasts of weather at the edges of the domain or far forward in time are likely to be erroneous. Reliable forecasts beyond 48 hours are not possible with continental-scale limited domains.

10.2.3. GRIDPOINT AND SPECTRAL MODELS. A gridpoint model is the simplest type of model. Data is mapped to an array of gridpoints, and model equations are applied to these gridpoints to produce a forecast map. The other type of model is a spectral model. Weather data is mapped out as a series of mathematical waves. Model equations are applied to these waves, and then this configuration is transformed back to geographic coordinates to produce a forecast map. Currently the gridpoint model is in most widespread use.

10.2.4. VERTICAL COORDINATES. The performance of a model also depends on the type of vertical coordinate it uses. The most familiar coordinate for forecasters is the pressure coordinate.

Meteorological cancer

Forecasters are relinquishing their meteorological input into the operational product going to the user. Forecasters are operating more as communicators and less as meteorologists. Since this practice is increasing slowly with time, it can be called "meteorological cancer". By this is meant that today's forecaster can, if he chooses, and many do, accept numerical prognoses and guidance, put this into words, and go home. Not once does he have to use his meteorological knowledge and experience. This type of practice is taking place more and more across the United States, and it will be made easier to do with [computer technology].

LEONARD W. SNELLMAN
"Operational Forecasting Using Automated Guidance," 1977

The birth of numerical forecasting

We propose to build ... a fully automatic, electronic, digital machine with the following characteristics ...

* multiplication time of 0.1 to 0.2 milliseconds [about 700 FLOPS; in 2011 consumer graphics cards were widely available with speeds of over 1 trillion FLOPS]

* an electronic memory of about 4,000 numbers, 36 to 40 binary digits each [about 18 kB; most cell phones available in 2011 had thousands of times this much memory]

Among the fields which we intend to study ... the one of dynamic meteorology is among the most important."

PROPOSAL TO ESTABLISH THE "METEOROLOGY PROJECT"
Frank Aydelotts, Princeton, N.J.
Institute for Advanced Study
May 8, 1946

This system is often used by simple primitive equation models. Isentropic coordinates may also be used.

However, horizontal and isentropic coordinates cause problems because those surfaces tend to intersect terrain, making it impossible to produce a calculation at the interface. A favored workaround is to use sigma coordinates which follow the terrain, so that $\sigma=0$ in the vacuum of space and $\sigma=1$ at the Earth's surface. But this in turn has its own problems; for example, sigma coordinates do not allow for accurate calculation of pressure gradient forces on sloped surfaces.

Because of this, many models, including the RUC, GFS, ECMWF, and NCEP WRF, use a hybrid coordinate system which makes use of both methods, with sigma coordinates below roughly 400 mb and pressure coordinates above that level. It eliminates many of the problems inherent in both models, though the models may have problems at the junctions between the two coordinate systems.

One unusual method worth noting is the eta scheme, which handles the lowest levels throughout the domain as a series of steps. This was the basis for the Eta model, produced by NCEP from 1993 to 2006. The scheme was known for its excellent ability to handle cold air damming situations and lee side cyclogenesis.

10.3. Numerical forecast production

There are four steps in the prediction process of a dynamic model: analysis, initialization, prediction, and post-processing.

10.3.1. ANALYSIS. In the first step of the analysis, the computer sorts through weather observations and maps the temperature, moisture, and pressure fields to a series of geographical grids. Some sophisticated methods have been developed for doing this. One of the most favored techniques is called "optimum interpolation", a three-dimensional method which considers the quality of the data and statistical relationships among the variables. Most of the models also rely on "first guess" fields, which uses the output of previous model runs as a starting point and adjusts it with new observations. This significantly improves the accuracy of the run and results in fields which are better balanced..

10.3.2. INITIALIZATION. The computer model thinks that the atmosphere is following equations. Therefore the initialization process is designed to adjust the analysis to be compatible with the prediction equations. Mathematical noise is removed from the fields, which fine-tunes the initial fields. Without this step, the prediction equations would produce unexpected errors, causing the model to "blow up".

10.3.3. PREDICTION. The primitive equations of motion and other equations derived from them, along with empirical relationships for friction, evaporation, precipitation, and so forth, are applied to the initial fields. This produces a prediction for a few minutes into the future. The computer repeatedly runs these equations, stepping forward a few forecast minutes at a time and producing new data fields. This is done until the run has gone as far into the future as requirements dictate. Statistical forecasts tend to generate the forecast directly without these intermediate steps.

10.3.4. POST-PROCESS. Once the equations for a particular time are solved, the model output is interpolated from the model coordinates to the display coordinates. It may also be filtered to account for biases and deficiencies in the model. The end results are what we see on the model output charts.

10.3.5. ENSEMBLE FORECASTING. Computing power increased enormously during the 1990s, allowing the development of ensembles. This is the technique of creating multiple model runs, each with a slightly different model package or a slightly different initialization. By doing this, it is possible to get probabilistic information on the model output, seeing where models are confident and where they are not. One ensemble system run at NCEP is SREF (Short-Range Ensemble Forecast system) which creates different initial conditions in each ensemble member.

10.3.6. BAYESIAN AVERAGING. Another scheme that takes advantage of the ability to run multiple models is the assessment of each ensemble member against actual weather to rank each one by their skill. When a new ensemble forecast is produced, a composite chart is created from all members, assigning the highest weight to members which have shown the best performance in previous runs and lowest weight to the worst performers. In this way, forecast charts are created using primarily the best available ensemble members.

In spite of progress made in the development of quantitative [computer] forecast techniques, the conventional forecaster will have an important part to play. His wide experience of local and regional conditions, orographic and topographic influences, moisture and pollution sources, etc, will be invaluable in supplementing the machine-made forecasts. While the machines provide the answers that can be computed routinely, the forecaster will have the opportunity to concentrate on the problems which can be solved only by resort to scientific insight and experience. Furthermore, since the machine-made forecasts are derived, at least in part, from idealized models, there will always be an unexplained residual which invites study. It is important, therefore, that the forecaster be conversant with the underlying theories, assumptions, and models. In particular, it is important that he be able to identify the "abnormal situation" when the idealized models (be they dynamical or statistical) are likely to be inadequate.

SVERRE PETTERSSEN,
"Weather Analysis and Forecasting," 1956

Up until the 1950s, weather forecasting in the United States was truly guesswork, relying upon rules of thumb, conceptual models, and experience. It was in 1958 however that the National Meteorological Center succeeded in producing a stable operational numerical forecast and began transmitting routine 500 mb forecasts to end users over facsimile circuits. This was essentially output from a single-layer barotropic model. A major improvement arrived in 1968 with the 6LPE (6-layer primitive equation model), forming the basis for the suite of numerical forecast products that forecasters are familiar with today. Late in 1971 forecasters began receiving the LFM (Limited-area Fine Mesh) model, which featured a 160 km grid for North America and was run out to 48 hours. Then in 1985 the Nested Grid Model (NGM) arrived, containing an 80 km grid overlaid on the LFM domain. It was known as the Regional Analysis and Forecast System (RAFS). The NGM model continued to be used until 2009.

Although the Air Force had already been running global models for some time, NOAA introduced the Global Spectral Model (GSM) in August 1980, a 30-wave spectral model. It was a framework for the Aviation (AVN) system, focusing on quick assimilation and quick output, and the Medium Range Forecast (MRF) system which was geared toward meteorological work beyond 3 days in the future. The GSM was eventually increased to 80 waves by 1990, allowing it to resolve finer detail around the world. This model is now known as the GFS (Global Forecast System) and has been greatly improved since then.

The next major milestone at NCEP was the introduction of the Eta model (named for the Greek-letter, not an acronym) in 1993. This model was better at handling interfaces between vertical coordinates and terrain than previous models, and the improvements in forecasts of weather systems around the Appalachians and the Rockies were noteworthy.

10.4. An overview of available models

There are various types of models that are in national or international use, which are described here. Many other types of models exist, but in general they are available only to a limited number of users and may never be encountered in day-to-day forecasting.

10.4.1. Weather Research and Forecasting (WRF) Model. Perhaps the most significant modelling milestone in many decades was the rollout of the WRF in 2000. The WRF is a highly flexible, modular *modelling system* designed to provide a unified framework for both the operational and research communities. Its roots are traced back to the MM5 model, developed by Penn State University in the 1970s. The WRF is robust enough that it was adopted by NCEP in 2005.

As it is a modelling system, it does not have any specific configuration of horizontal and vertical resolution or physics packages. These are established by the end user. However, two major flavors of the WRF exist: the ARW core and the NMM core (Advanced Research WRF and Nonhydrostatic Mesoscale Model, respectively). The ARW core is a hydrostatic model developed by NCAR and is favored for research work. The NMM core was developed at NCEP and is geared more toward operational forecasting.

Thanks to the WRF's portability, users may compile and run the WRF on any computer with Linux. Though this gives some semblance of "personalizing" the model run, nearly all off-the-shelf WRF packages rely on the same sets of analysis fields from NCEP. This means that any biases in the model initialization will propagate in the same way to most WRF users and result in similar output.

10.4.2. NCEP North American Model (NAM). The NAM describes a custom configuration of the WRF-NMM model outlined above. The model consisted of the Eta model up until 2006, when it was replaced with the WRF-NMM. It uses a resolution of 12 km with 60 sigma-pressure hybrid surfaces. NCEP also produces several High Resolution Window (HRW) products which are subcontinental scale: two covering the U.S. and a few others on Alaska, Hawaii, and Puerto Rico.

10.4.3. NCEP Global Forecast System (GFS). The GFS is an outgrowth of the GSM model initially developed in 1980. It is a spectral model which as of 2011 solved 574 waves on 64 sigma-pressure hybrid levels and ran its forecasts 16 days into the future.

10.4.4. NCEP RAPID UPDATE CYCLE (RUC) MODEL. The RUC model was introduced in 1994. It was designed to provide a forecast model that could run on an hourly basis, allowing other data besides radiosonde and surface observations to be ingested. This was an important capability that addressed aviation and severe weather interests. As of 2011 the RUC was a gridpoint model with 13 km resolution and 50 hybrid sigma-isentropic layers. The RUC is gradually being replaced by the Rapid Refresh model (below). Output from the RUC can be viewed at <ruc.noaa.gov>.

10.4.5. NCEP RAPID REFRESH MODEL (RR). The Rapid Refresh model, introduced in 2010, is an outgrowth of the RUC model which combines the physics packages of the WRF-ARW and RUC models. It integrates the radar assimilation capabilities of the RUC. By 2017 the Rapid Refresh is expected to be expanded to cover all of the globe. The High Resolution Rapid Refresh (HRRR) is a 3 km model that was originally part of the RUC, but is being transitioned to the RR model. This system can be viewed at <rapidrefresh.noaa.gov>.

10.4.6. ECMWF MODEL. Since the 1970s, the European Centre for Medium Weather Forecasting has operated a very sophisticated global modelling system. Unfortunately as with many European weather initiatives much of its output has been intentionally kept out of the hands of the general public, impeding its acceptance in operational forecasting outside of government forecast offices. Some basic fields from the model are now available on the Internet, however. As of 2011 it was a spectral model resolving 1279 waves in 91 hybrid sigma-pressure layers.

10.5. Limitations of models

It is extremely important to remember that models are only simulations of the atmosphere using current data. They are never 100% correct, and in unusual situations (e.g. major storms) significant errors may be seen. Here are some factors that contribute to the inaccuracy of numerical models.

10.5.1. SMALL AMOUNT OF OBSERVATIONAL DATA. A model might have great resolution — 30 or 40 miles, but useful observations may be 200 to 300 miles apart. Assuming the observations are

Receptivity allows one's thoughts to balance properly the daily stream of observational data, model data, and environmental clues, while enjoying the process of determining which clues are most critical. The more you love the forecasting and ingestion of data and environmental clues, the greater the success you will have. Get lost in the joy of the process!

ALAN R. MOLLER
NWS Forecaster, 2001

accurate, we can say that the more inaccurate the initial analysis is, the greater the likelihood that forecast charts will contain inaccuracies, especially as the computer steps forward in time.

10.5.2. BOUNDARY ERRORS. A global-domain model such the GFS is immune to boundary errors since the model encompasses all points of the globe. However, limited-domain models such as the basic WRF only cover a limited area. Therefore the edges of the domain gradually become unrepresentative with time (the model does not account for real-world systems outside the boundary of the model). On the ETA run, for example, significant errors are seen in the central Pacific at the 48 hour point due to boundary errors.

10.5.3. LOSS OF MESOSCALE FEATURES. Grid spacing often precludes detection and prediction of some subsynoptic features. There are many short waves in the lower and middle troposphere that are only a few degrees or less wide in longitude. Since each one is baroclinic, each short wave has its own area of upward and downward vertical motion. The model may be unable to forecast it and the weather it creates.

10.5.4. PARAMETERIZATION OF PHYSICAL PROCESSES. Processes such as solar radiation, terrestrial radiation (including radiation reflected and re-reflected by cloud tops, bases, etc), evaporation from below clouds, precipitation, and other processes must be explicitly or implicitly accounted for in order to have the model produce a realistic simulation of the atmosphere. Clouds and precipitation are some of the most important physical processes a model must predict. Unfortunately these processes may not be adequately accounted for in the model, or in simple models, some processes may be ignored.

10.5.5. PARAMETERIZATION OF CONVECTION. Most operational forecast models handle convection unrealistically. There is inadequate computer power to resolve these processes, many of which are not fully understood. Furthermore, if a major thunderstorm is occurring at the time of a sounding and this sounding is inadvertently entered into the analysis of a model, the model could blow the situation up into a hurricane-like storm (this is called "convective feedback"). The forecast models of the Air Force and Navy are aimed more at cloud prediction because of obvious aviation concerns, therefore they include stronger attempts at parameterizing for convection.

10.6. Climatological patterns

Climatological patterns are well beyond the scope of this book since they have scales on the order of months or years, and meteorology mostly concerns itself with time scales of hours and days. Due to the prominence of climate in news headlines and their ability to shape the large-scale weather pattern, they deserve brief mention.

10.6.1. EL NIÑO. The so-called El Niño, in its simplest terms, is an episode where the ocean temperatures in the East Pacific are warmer than normal. This is the warm phase of the El Niño Southern Oscillation (ENSO). It is generally caused by a weaker east-to-west component in the tropical flow across the Pacific, which causes weaker upwelling of cold ocean waters along the coast of the Americas. The episode lasts a few months to a few years, and may occur every few years. El Nino's primary effect on day-to-day meteorology is to impart more thermal energy to weather systems in the East Pacific, creating an active storm track into California. The southern United States and Gulf Coast area experience wet conditions, while the northern U.S. experiences dry weather. The pattern also tends to suppress hurricane activity in the Atlantic due to somewhat stronger westerlies bringing stronger deep-layer shear.

10.6.2. LA NIÑA. The La Niña pattern is essentially the opposite of an ENSO episode. This is the cold phase of the El Niño Southern Oscillation (ENSO). The Eastern Pacific is colder than normal, largely because of strong east-to-west tropical flow that promotes upwelling of cold ocean waters. This pattern tends to produce dry, stagnant weather in the southern and central United States, while the northern U.S. and Canada experience above-normal precipitation. Shear tends to be weaker in the Atlantic due to weaker westerlies through this area, which tends to enhance the incidence of hurricanes.

10.6.3. TELECONNECTIONS. Climatologists monitor the mean pressure difference between two stations in different parts of a continent in order to gauge the potential distribution of energy.

* *Pacific North American (PNA).* This measures energy across the North Pacific basin. In a positive [negative] phase, the northerly temperate Pacific has unusually low [high] pressure while the southerly subtropical Pacific has unusually high [low] pressure. The result is stronger [weaker] westerlies through the Pacific region. A +PNA favors zonal patterns in

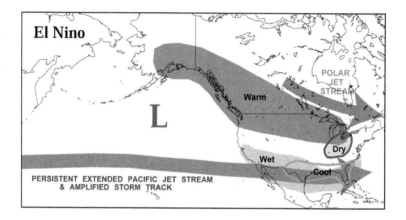

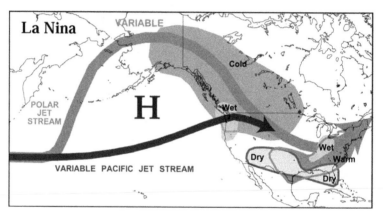

the Pacific, while a –PNA pattern favors blocking and often results in severe cold outbreaks into the central United States as cutoff highs build into Alaska.

* *North Atlantic Oscillation (NAO).* This pattern measures energy across the North Atlantic. It is similar to the PNA pattern in that a +NAO mode favors unusually low pressure over the northerly subarctic region of the Atlantic and high pressure over the southerly temperate latitudes of the Atlantic, producing strong westerlies. Likewise, the –NAO pattern has an opposite effect, resulting in weak westerlies and blocking.

* *Arctic Oscillation (AO).* The AO index looks at pressure patterns between the arctic and temperate latitudes. In a +AO [–AO] phase, the arctic has unusually low [high] pressure, and favors a strong [weak] Icelandic Low while the Aleutian Low is weaker [stronger] than normal. In the United States, a +AO pattern produces warm weather across the Midwest and wetter conditions in the west, while a –AO

pattern brings cold weather to much of the country and
drier conditions in the Desert Southwest.

Chapter Ten
REVIEW QUESTIONS

1. How might a forecaster perform a thorough analysis but not a proper diagnosis?

2. What is a shortfall of using a decision tree to make a forecast?

3. Why was numerical modelling not possible in the early 20th century?

4. What type of model is an analog and how does it work?

5. Explain how ensemble forecasting works.

6. What makes a spectral model different from a gridpoint model?

7. Why does Bayesian averaging give more accurate model output?

8. Meteorological calculations fail if any of the gridpoints surrounding a given gridpoint intersects the terrain. What is a method that can be used to avoid this?

9. Name the highly modular community-based dynamical model which is in widespread use today.

10. Explain the basic cause of an El Nino cold episode and what effect it has on the west coast of North America.

APPENDIX

Forecaster's guide to cloud types

The full definition of cloud types are published in the World Meteorological Organization's *International Cloud Atlas*, initially published in 1896 and remaining in print ever since then. Presented here, however, is a minature cloud atlas that provides a brief definition of the WMO cloud type and then discusses the cloud in a forecasting context. It should be noted that ten photos is not enough to convey all the possible appearances of a cloud type, so readers who are interested in this subject should try to obtain a copy of the above World Meteorological Organization book. Some good alternate selections include the *National Audubon Society Field Guide to North American Weather*, the *Peterson First Guide to Clouds and Weather*, Jay Pasachoff's *Field Guide to the Atmosphere*, Richard Scorer & Arjen Verkaik's *Spacious Skies*, and Richard Scorer's *Clouds of the World* (the latter two are out of print).

Cumulus consists of clearly detached, dense clouds with sharp outlines, developing vertically in the form of rising mounds, domes, or towers. The bulging upper part often resembles a cauliflower. The sunlit parts of these clouds are mostly brilliant white; their base is relatively dark and nearly horizontal.

They are indicative of a convectively unstable atmosphere. The clouds are caused either by solar heating or conductive heating at the Earth's surface or from ascent of moist, elevated layers aloft. In either case, convective updrafts are produced through a relatively shallow layer, generally less than 2 or 3 km deep. These thermals ascend and reaching their condensation level. The clouds seen here are towering cumulus in East Texas just ahead of Hurricane Rita in 2005.

Cumulus

Cumulonimbus

Cumulonimbus is a heavy and dense cloud, of great vertical extent, in the form of a mountain or a huge tower. By convention, it is the only cloud that can produce thunder, lightning, and/or hail, but it doesn't necessarily do so. At least part of its upper portion is glaciated (smooth, fibrous, or striated), and it often spreads out in the shape of an anvil or vast plume. Under the base of this cloud, which is often very dark, there are frequently low ragged clouds either merged with it or not, and precipitation, sometimes in the form of virga in dry air. Aloft, there is the horizontal spreading of the highest part of the cloud, leading to the formation of an "anvil". Often, strong upper-level winds blow the anvil downwind in the shape of a half anvil or vast plume.

From a forecasting standpoint, cumulonimbus are indicative of moist, buoyant parcels which are ascending and have reached the free atmosphere. Widespread release of latent heat and production of ice crystals in the upper portions of the cloud are occurring.

Stratocumulus is a grey or whitish patch, sheet, or layer of cloud which almost always has dark parts, composed of tessellations, rounded masses, or rolls, which are non-fibrous (except for virga), and which may or may not be merged. Stratocumulus may sometimes be confused with altocumulus. In very cold weather, stratocumulus may produce abundant ice crystal virga.

Stratocumulus indicates the ascent of convective updrafts in a weakly unstable atmosphere through a shallow layer, generally less than 1 or 2 km deep. They are often produced by the spreading of cumulus clouds beneath an inversion beneath 1 or 2 km AGL. Spreading at higher levels generally results in altocumulus.

Stratocumulus

Altocumulus

Altocumulus is a white or grey patch, sheet, or layer of cloud, generally with shading, and composed of laminae, rounded masses, rolls, etc. which are sometimes partially fibrous or diffuse and which may or may not be merged. Most of the regularly arranged small elements usually have a visual width or between 1 and 5 degrees. Altocumulus sometimes produces descending trails of fibrous appearance (virga).

Altocumulus generally indicates the presence of weak convective updrafts in the middle troposphere. This is generally associated with the presence of high relative humidities and weak instability in the middle troposphere.

Cirrocumulus is a thin, white patch, sheet, or layer of cloud without shading, composed of very small elements in the form of grains, ripples, etc. merged or separate, and more or less regularly arranged. Most of the elements have a visual width of less than 1 degree. A rare cloud, cirrocumulus is rippled and is subdivided into very small cloudlets without any shading. It can include parts which are fibrous or silky in appearance but these do not collectively constitute its greater part.

Much like altocumulus, cirrocumulus indicates the presence of weak convective updrafts in the upper troposphere.

Cirrocumulus

Cirrus

Cirrus is made up of detached clouds in the form of white delicate filaments, or white or mostly white patches or narrow bands. These clouds have a fibrous appearance, or a silky sheen, or both. Cirrus is whiter than any other cloud in the same part of the sky. With the sun on the horizon, it remains white, while other clouds are tinted yellow or orange, but as the sun sinks below the horizon the cirrus takes on these colors, too, and the lower clouds become dark and/or grey. The reverse is true at dawn when the cirrus is the first to show coloration.

From a forecasting standpoint, cirrus indicates the presence of moisture in the upper troposphere, with conditions at that level not favorable for the variations known as cirrostratus and cirrocumulus.

Cirrostratus is a broad, transparent whitish cloud or veil of fibrous or smooth appearance, totally or partially covering the sky, and generally producing halo phenomena. The cloud usually forms a veil of great horizontal extent, without structure and of a diffuse general appearance. It can be so thin that the presence of a halo may be the only indication of its existence.

Cirrostratus generally indicates a broad region of forced ascent in the upper troposphere, with this layer having relatively weak lapse rates.

Cirrostratus

Altostratus is a greyish or bluish cloud sheet or layer of striated, fibrous, or uniform appearance, totally or partially covering the sky, and having parts thin enough to reveal the sun at least vaguely, as if through ground glass. Altostratus prevents objects on the ground from casting shadows. If the presence of the sun or moon can be detected, this indicates altostratus rather than nimbostratus. If it is very thick and dark, differences in thickness may cause relatively light patches between darker parts, but the surface never shows real relief, and the striated or fibrous structure is always seen in the body of the cloud. At night, if there is any doubt as to whether it is altostratus or nimbostratus when no rain or snow is falling, then, by convention, it is called altostratus. Altostratus is never white, as thin stratus may be when viewed more or less towards the sun. *(Tim Vasquez)*

Altostratus is normally the result of when large areas of isentropic ascent develop in an air mass with high relative humidity.

Altostratus

Nimbostratus is a broad, featureless grey cloud layer, thick enough to obscure the sun and whose appearance is diffused by rain, snow, or extensive virga. It may be formed from the thickening of altostratus or stratocumulus. The base is frequently partially or totally hidden by ragged scud clouds (stratus). Care must be taken not to confuse these with the base of the nimbostratus. Nimbostratus can be distinguished from thick stratus by the type of precipitation it produces. If hail, thunder, lightning, or showery weather are produced by the cloud, it is then classified as cumulonimbus. In this image, the cloud's textures are caused by virga falling from the cloud. (Tim Vasquez)

From a forecasting standpoint, nimbostratus is the culmination of strong, persistent forcing of a humid air mass across a large region. Once in place, nimbostratus and stratus layers tend to persist until dry air and subsidence move into the region.

Nimbostratus

Stratus is an amorphous, very low cloud, with a fairly uniform base. It may precipitate drizzle, snow, or snow grains. Rain may be present, caused by other clouds, and such "bad weather stratus" is known as pannus or scud. When the sun is visible through stratus, its outline is clearly discernible.

Stratus is generally the result of widespread ascent in the lower troposphere in the presence of moderate to high static stability.

Stratus

APPENDIX 2
Surface station plots

Station plots show the weather at specific locations. The station is marked with a hollow circle if the observation was taken by radiosonde, or a square if from an automated observation. The size of the square or circle is about 1/8th inch (2 to 4 mm). If the station is not capable of transmitting sky condition data, then a + mark may be used for the station mark.

The circle is shaded according to the total cloud cover present. A clear circle indicates clear skies, and a shaded circle indicates cloudy skies. An "X" in the circle indicates the sky is obscured by precipitation or other weather.

From the station marking, a wind shaft is drawn to depict wind direction, with barbs for every 10 kt of wind and half barbs every 5 kt. If a gust is reported, it is written near the barbs as "Gxx" where xx is the wind gust value.

At the top left and bottom left are temperature and dewpoint value, respectively, in either degrees Fahrenheit or degrees Celsius.

At the top right, sea-level pressure is written in three digits, using the tens, units, and tenths places of the value in millibars. The analyst can instead plot altimeter setting (QNH), in which case the units, tenths, and hundredths place is plotted in inches of mercury.

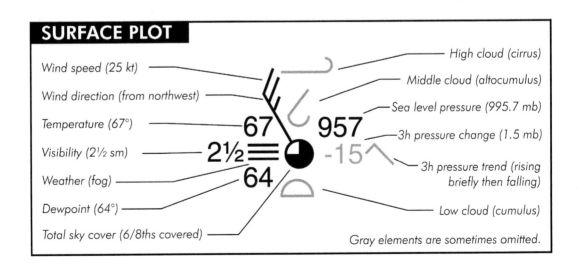

SURFACE PLOT

Wind speed (25 kt)

Wind direction (from northwest)

Temperature (67°)

Visibility (2½ sm)

Weather (fog)

Dewpoint (64°)

Total sky cover (6/8ths covered)

High cloud (cirrus)

Middle cloud (altocumulus)

Sea level pressure (995.7 mb)

3h pressure change (1.5 mb)

3h pressure trend (rising briefly then falling)

Low cloud (cumulus)

Gray elements are sometimes omitted.

67   957

2½   -15

64

APPENDIX 3
Surface chart analysis procedures

Surface chart analysis is covered in detail in the chapter on Surface Analysis.

Some possible isopleths are listed below.

Isopleth	Quantity	Base value	Interval	Color
Isobars	Pressure	1000 mb	2 or 4 mb	Solid black
Isotherms	Temperature	0 °F 0 °C	2 or 5 F° 2 or 5 C°	Red
Isodrosotherms	Dewpoint	0 °F 0 °C	2 or 5 F° 2 or 5 C°	Green
Isohume	Mixing ratio	10 g/kg	1 or 2 g/kg	Green
Isotach	Wind speed	0 kt	5 or 10 kt	Purple
Streamline	Wind streamfunction	-		Black

APPENDIX 4
Upper air station plots

Station plots show the weather at specific locations. The station is marked with a hollow circle if the observation was taken by radiosonde, a square if from an aircraft observation, or no station marking if it's a satellite observation. The size of the square or circle is about 1/8th inch (2 to 4 mm).

By convention, this station marking is shaded if the dewpoint depression is 5 Celsius degrees or less. This suggests the presence of clouds at that level.

From the station marking, a wind shaft is drawn to depict wind direction, with barbs for every 10 kt of wind and half barbs every 5 kt.

At the top left and bottom left are temperature and dewpoint depression, respectively, in degrees Celsius. In specific situations, a forecaster may instead write

the actual dewpoint value here, but if this is done an annotation should be clearly marked on the chart to alert readers.

At the top right is the geopotential height in meters, for which certain decimal places are dropped (see illustration). The lower right may sometimes contain a 12-hour height change value, which is commonly italicized if plotted with a computer program.

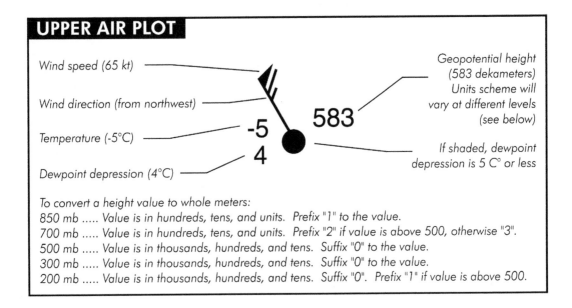

UPPER AIR PLOT

Wind speed (65 kt)

Wind direction (from northwest)

Temperature (-5°C)

Dewpoint depression (4°C)

-5
4

583

Geopotential height (583 dekameters) Units scheme will vary at different levels (see below)

If shaded, dewpoint depression is 5 C° or less

To convert a height value to whole meters:
850 mb ..... Value is in hundreds, tens, and units. Prefix "1" to the value.
700 mb ..... Value is in hundreds, tens, and units. Prefix "2" if value is above 500, otherwise "3".
500 mb ..... Value is in thousands, hundreds, and tens. Suffix "0" to the value.
300 mb ..... Value is in thousands, hundreds, and tens. Suffix "0" to the value.
200 mb ..... Value is in thousands, hundreds, and tens. Suffix "0". Prefix "1" if value is above 500.

APPENDIX 5
Upper air chart analysis procedures

Upper Tropospheric Charts

Forecasters refer to upper tropospheric charts to track the jet stream, take note of the steering flow, and examine other upper-level features. The ideal upper tropospheric chart to use is the one that is in the highest levels of the troposphere but not in the stratosphere itself. Since the tropopause height varies with the seasons, forecasters in the United States traditionally use the 300 mb (~30,000 ft) chart during the winter and the 200 mb (~39,000 ft) chart during the summer. Some forecasters use the 250 mb (~34,000 ft) during spring and fall.

Features typically analyzed are contours and isotachs. The axis of jet streams are outlined with a red pen. The forecaster looks at the large waves and ridges, which outlines the long wave pattern (to be discussed in upcoming chapters) and gives an idea of the three-dimensional thermal structure across a continent.

Middle Tropospheric Charts

Mid-tropospheric charts are useful because they are close to the level of non-divergence, which in terms of mass is near the middle of the troposphere and is generally a level where synoptic-scale ascent and descent are taking place rather than convergence and divergence. The favored mid-tropospheric chart is the 500 mb (~18,000 ft or 5 km) chart. It shows the broad-scale flow and jet streams seen on upper-level charts, yet it reveals some of the intricate details of low-level storm systems.

Features analyzed include contours and isotachs. The axis of the jet stream is outlined in thick red pen. Short wave troughs (boundaries marked by strong cyclonic wind shifts) are outlined using a solid black line. Short wave ridges (features marked by strong anticyclonic wind shifts) are outlined using a black sawtooth line.

Absolute vorticity patterns, which will be discussed later, cannot be created by hand but are frequently added to 500 mb charts through computer analysis at an interval of 2×10^{-5} s^{-1}. If vorticity patterns are available, forecasters will adjust and refine the positions of short waves using other tools and imagery. Areas of strong positive vorticity advection (PVA) are shaded in red, and areas of strong negative vorticity advection (NVA) are shaded in blue.

The 700 mb chart is sometimes neglected since it shows similar patterns to the 500 mb chart. It is still useful, though, since many short wave disturbances are found at this level too. The 700 mb level winds can also provide a quick estimate of any thunderstorm movement. During severe weather events, the 700 mb temperature patterns are critical since they indicate the strength and extent of the mid-level cap, a feature that suppresses thunderstorm activity or delays it until the instability is dramatically higher.

Lower Tropospheric Chart

The 850 mb (~5000 ft or 1.5 km) chart helps complete the three-dimensional picture of tropospheric weather. It is best used as an upward extension of the surface chart, with differences indicating the presence of shallow air masses or a change from the boundary layer to the free atmosphere. It must be remembered, however, that the 850 mb chart is at about 5000 ft and in higher terrain it may be near the surface or below ground.

With the 850 mb chart, boundaries and fronts can be identified and located. Low-level warm air advection is associated with upward vertical motion, so areas where isotherms cross the height contours often indicate potential areas of cloud and precipitation development. Fronts aloft are also located. These fronts, in accordance with basic rules about baroclinic zones, are drawn on the warm side of strong temperature gradients.

Moisture is also outlined on the chart. For typical forecasting that does not involve severe weather, areas of dewpoint depressions 5 C deg or less are outlined in thick green, signifying areas of possible clouds. For severe weather forecasting, contours for dewpoint temperature are drawn every 5 C deg. Since upper-air station plots, by convention, indicate dewpoint depression, the analyst may need to use quick arithmetic on each station plot to determine its dewpoint temperature.

Upper air chart standards
Conventions for upper-level charts are as follows:

Level	Height (Standard Atmosphere)	Contour Base	Contour Interval
100 mb	53,083 ft / 16,180 m	1608 dam	12 dam
200 mb	38,662 ft / 11,784 m	1176 dam	12 dam
250 mb	33,999 ft / 10,363 m	1008 dam	12 dam
300 mb	30,065 ft / 9164 m	900 dam	12 dam
400 mb	23,574 ft / 7185 m	720 dam	8 dam
500 mb	18,289 ft / 5574 m	558 dam	6 dam
600 mb	13,801 ft / 4206 m	420 dam	6 dam
700 mb	9882 ft / 3012 m	300 dam	3 dam
850 mb	4781 ft / 1457 m	150 dam	3 dam
925 mb	2500 ft / 762 m	750 m	30 m

APPENDIX 6
An isoplething tutorial

Those who are new to forecasting often have difficulties with hand analysis, especially the technique of isoplething.

Shown here is a sample field of temperatures across the eastern United States. We will analyze isotherms, which are lines of equal temperature, at an interval of every 5 degrees. Examining the temperature ranges across the chart we can see that we will be plotting lines at 15, 20, 25, 30, 35, 40, 45, 50, 55, 60, and 65 degrees.

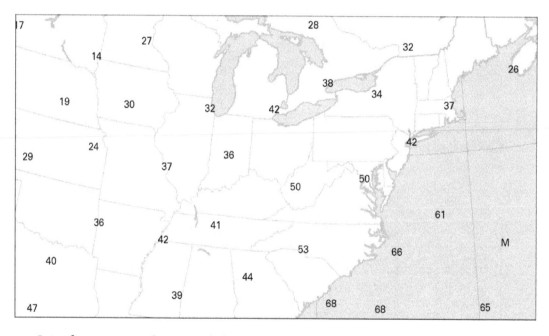

It is often easiest to plot an isopleth in the middle of this range. We will select the value of 50 degrees. At this point you may attempt to plot the 50-degree isopleth or turn the page for detailed instructions on how to plot this line.

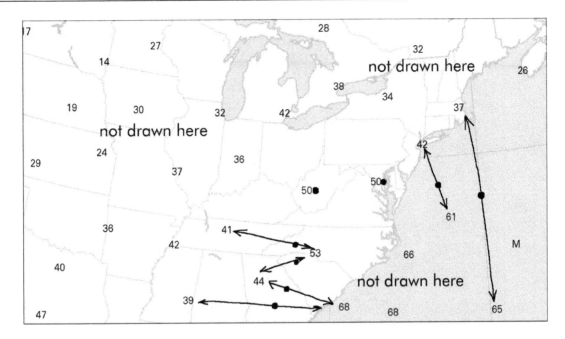

The first task is to mentally outline the areas where the line will be drawn. This is not done all at one; rather, start at the chart edge where the line should appear, and look ahead a couple of inches to find where it will lead. Once the *general path* it will take is determined, begin looking at the closest stations where the line will pass between (arrows seen here). By estimating where the line should fall between each station, this will help clarify its *immediate path*. Here we will start in Georgia at the bottom of the chart and work northward.

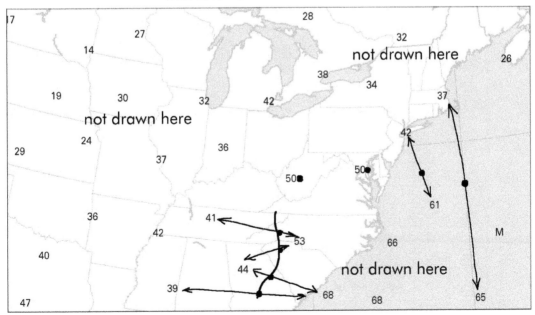

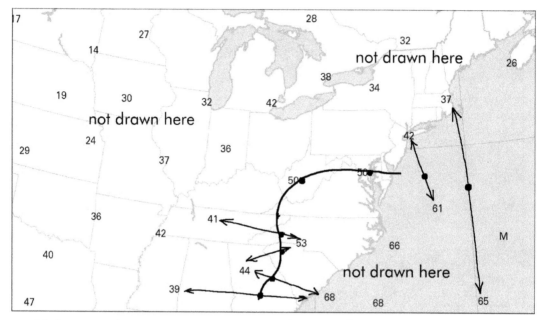

It's fortunate when two stations are reporting the exact value we are looking for, such as Charleston WV and Washington DC, as shown here. But instead of drawing a simple straight line, we take into consideration the *general path* and form somewhat of a curve as shown here.

In the image below, note that the isopleth stops before reaching the right edge of the chart. The line should stop at this point because we don't have sufficient data to draw it any further, and the isopleth should reflect the actual data field.

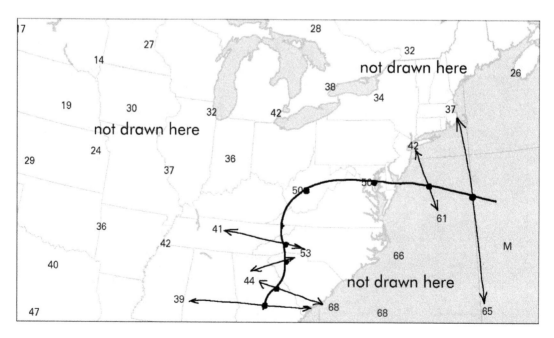

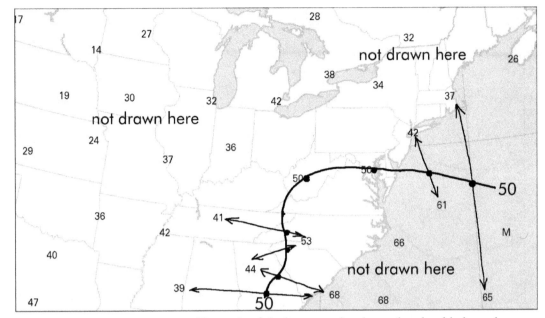

The ends of the line near the chart edge should always be labelled with the isopleth value. If the lines do not reach the edge, appearing as a "closed isopleth" as might be seen concentric isobars around a surface low, the label is placed at the top of the closed isopleth

An easy way to draw more isopleths is to not go to the next incremental value (e.g. 45 or 55) but skip to the second or fourth isopleth away. This would be a value of 40, 30, or 60. This makes it easy to see the general paths at play and helps plot the ispleths in between. Now we will plot the 30 degree isopleth.

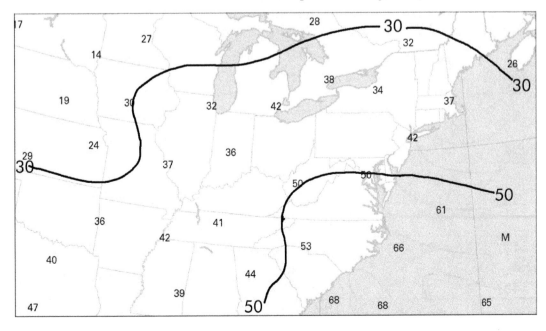

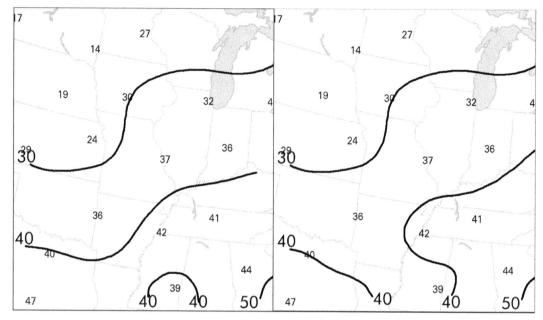

The 40 degree isopleth is now very easy to find because most parts should lie about halfway between the 30 and 50 degree lines. Note that there are at least two possible solutions for this line in Mississippi, where an outlying 39-degree temperature exists. The temperatures seem to be below 40 on a northwest-southeast axis but above 40 on a northeast-southwest axis. Both solutions are valid.

When all lines are drawn, the resulting isotherm field looks like the chart below. Double-check that no isopleths have been missed, that it conforms to the data, and all are properly labelled.

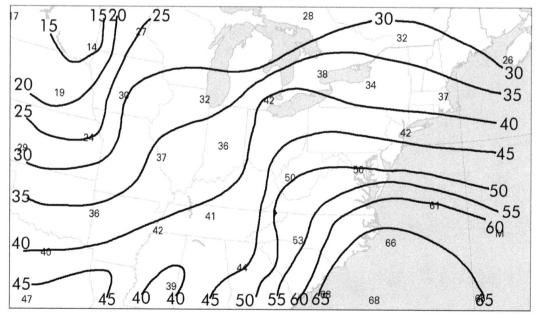

Additional advice for an accurate hand analysis . . .

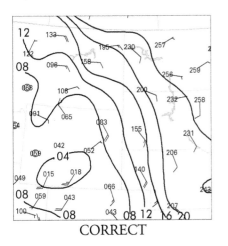

CORRECT

Do not smooth over a data point that does not seem to fit. If it is obviously erroneous, circle it and ignore it, but otherwise, force the isopleths to fit the data. The error here appears in northwest Wyoming.

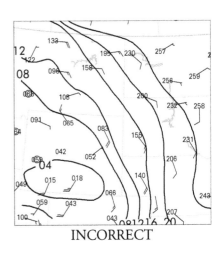

INCORRECT

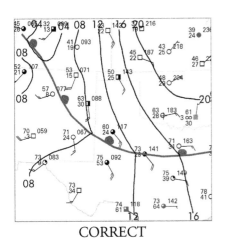

CORRECT

Isobars (pressure lines) will kink at frontal boundaries. This is a consequence of the change in density across the boundary. The kink should be added when the drawn isobars and frontal placements are finalized.

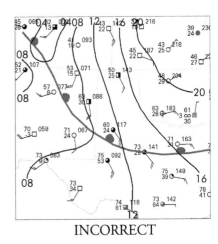

INCORRECT

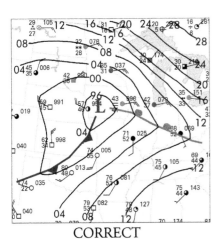

CORRECT

Instead of placing pressure centers directly on the most extreme pressure found on the chart, follow the winds at that station using Buys Ballot's Law to adjust the pressure center to the correct location.

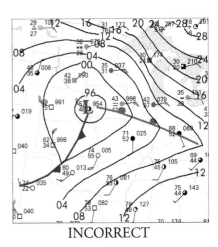

INCORRECT

APPENDIX 7
Conversions & symbols

Listed below are some common conversions used in forecasting. Also included is the Greek alphabet with some of the variables and constants they might represent.

SI UNITS

Unit	SI units	Abbv.	Nonstd equiv.
Distance	meter	m	3.28084 feet
Mass	kilogram	kg	2.2046 pounds
Time	second	s	—
Pressure	pascal	Pa	0.00001 bar
Energy	joule	J	10,000,000 ergs
Frequency	Hertz	Hz	1 cps (cycle per second)
Temperature	Kelvin	deg K	-273.16 deg Celsius

DISTANCE

Unit	Equals	Also equals	Also equals
ft	m x 0.3048	—	0.3048 m
m	ft x 3.2808	—	3.2808 ft
km	sm x 0.621	nm x 0.54	3280.8 ft
sm	km x 1.61	nm x 0.87	5280 ft
nm	km x 1.85	sm x 1.15	6080 ft

VELOCITY

Unit	Equals	Also equals	Also equals
m/s	km/h x 0.27	kt x 0.5144	mph x 0.4473
km/h	m/s x 3.6	kt x 1.85	mph x 1.609
kt	m/s x 1.944	km/h x 0.54	mph x 0.87
mph	m/s x 2.2356	km/h x 0.621	kt x 1.15

SCALES

Multiple	Prefix	Abbv.	Weather usage
1,000,000	mega-	M-	MHz (radio frequency)
1,000	kilo-	k-	km (distance); kg (mass)
100	hecto-	h-	hPa (press.; 1 hPa = 1 mb)
10	deka-	da-	dam (geopotential height)
1	—	—	—
0.1	deci-	d-	—
0.01	centi-	c-	cm (radar wavelength)
0.001	milli-	m-	mm (droplet size)
0.000001	micro	m-	mm (particle size)

Alpha	A	
	α	Angle
Beta	B	
	β	Rossby parameter
Gamma	Γ	Static stability
	γ	Lapse rate
Delta	Δ	Finite difference
	δ	
Epsilon	E	
	ε	Random error
Zeta	Z	
	ζ	Relative vorticity
Eta	H	
	η	Absolute vorticity
Theta	Θ	
	θ	Potential temperature; latitude
Iota	I	
	ι	
Kappa	K	
	κ	
Lambda	Λ	
	λ	Longitude
Mu	M	
	μ	Wave number; micron
Nu	N	
	ν	Frequency
Xi	Ξ	
	ξ	
Omicron	O	
	o	
Pi	Π	Product operator (math)
	π	3.14159; normalized pressure
Rho	P	
	ρ	Density
Sigma	Σ	Summation operator (math)
	σ	Sigma coordinate; static stability
Tau	T	
	τ	Time interval; friction force
Upsilon	Y	
	υ	
Phi	Φ	
	ϕ	Geopotential; phase lag
Psi	X	
	χ	Tendency of geopotential height
Psi	Ψ	
	ψ	Stream function for rot. velocity
Omega	Ω	Angular velocity of Earth
	ω	Vertical motion in p system

APPENDIX 8
Instability index summaries

CONVECTIVE AVAILABLE POTENTIAL ENERGY (CAPE), *B+*
The CAPE value is by far one of the most useful predictors for thunderstorms. It equals the vertically integrated positive buoyancy of an adiabatically rising parcel throughout the entire troposphere. In fact, updraft speed is closely tied to CAPE. Any positive CAPE value indicates that rising motion will occur, providing any lower inversions can be overcome. A values of up to 1000 J/kg indicates weak convection, 1000 to 2500 J/kg indicates moderate convection, and above 2500 J/kg indicates strong convection.

CONVECTIVE INHIBITION (CIN), CINH, B-
This value represents the negative energy area on a SKEW-T below the level of free convection, and is equivalent to negative CAPE. If natural processes fail to destabilize these negative areas, energy from forced lift will be required to move the negatively buoyant parcels upward to their LFC. Like CAPE, CIN is measured in J/kg. The significance of CIN depends on the buoyancy in layers below and the amount of available forced lift that can occur.

BRN SHEAR (BULK SHEAR)
This is simply the vector difference between the 0-6 km wind and the near-surface winds. Its chief use is to determine whether storms will be long-lived (if the downdraft will be separated from the updraft).

BULK RICHARDSON NUMBER (BRN)

This is defined by CAPE divided by the BRN Shear. This attempts to combine instability and bulk shear in order to separate storm types into various categories. BRN's less than 45 tend to support supercell structures (values below 10 tend to shear out any developing convection), while multicellular convection favors BRN's above 45. BRN by itself is a poor predictor of storm rotation, since it only addresses bulk shear (which determines a storm's longevity) and not low-level helicity (which allows for rotating storms). BRN yields the following results given a certain amount of instability and bulk shear:

Instability	Bulk Shear	BRN
Low	High	Low
Low	Low	Medium
High	High	Medium
High	Low	High

STORM-RELATIVE HELICITY (SRH)
This is a summation of streamwise vorticity through the storm inflow layer, and is dependent on figuring an appropriate value of storm motion (typically a value 30 degrees to the right of the 0-6 km wind with 75% of its magnitude, "30R75"). The SRH value is generally determined using computer diagnostics.

ENERGY-HELICITY INDEX (EHI)
This equals the product of positive SRH and CAPE, divided by 160,000. It is an attempt to combine the effects of CAPE in producing strong updrafts and storm-relative helicity in providing rotating

updrafts. EHI values above 1 indicate strong tornadoes, while violent tornadoes favor EHI values of 5 or greater.

K-Index (KI)

Equals $K = (T_{850} - T_{500}) + Td_{850} - Tdd_{850}$, where Td is dewpoint and Tdd is dewpoint depression. This represents thunderstorm potential as a function of vertical temperature lapse rate, low-level moisture content, and depth of the moist layer. It has been found that KI is good for forecasting heavy rain but not for determining whether storms will be severe. Values between 31 and 35 tend to favor scattered storm development, with higher values indicating numerous thunderstorms.

Lifted Index (LI)

This is simply the Celsius degree difference in temperature between a representative (mixed) low-level parcel lifted to 500 mb and the environmental 500 mb air. A parcel that is warmer than its environment has a negative LI. Values of -5 or less are associated with severe weather.

Showalter Index (SI)

This is the Celsius degree difference in temperature between an 850 mb parcel and the environmental 500 mb air. Since moisture depth is not taken into account, it is not a preferred method for figuring storm potential (the Lifted Index should be used). Values of -4 or less are associated with severe weather.

Severe weather threat index (SWEAT)

This combines low level moisture, instability, jet speeds, and warm advection. Values above 300 indicate a potential for severe storm development,

and above 400 indicate tornadic storms. The value is computed as:

$$SWEAT = 12(Td_{850}) + 20(TT-49) + 2(V_{850}) + V_{500} + 125(\sin(D_{500} - D_{850}) + 0.2)$$

where D is the wind direction and V is the wind speed

Total totals index (TT)

Combines lapse rate and low level moisture to estimate the potential for severe convection. When the TT exceeds 50, a few severe storms are indicated. Values above 52 favor scattered to numerous thunderstorms, and values above 56 support numerous thunderstorms with scattered tornadoes.

$$TT = (T_{850} - T_{500}) + (Td_{850} - T_{500})$$

Vertical totals index (VT)

Determines the mid-level lapse rate as a rough approximation of deep-layer static stability.

$$VT = T_{850} - T_{500}$$

Cross totals index (CT)

Relates low-level moisture to mid-level temperatures.

$$CT = Td_{850} - T_{500}$$

Wet-bulb zero (WBZ)

This indicator, generally used for hail prediction, equals the lowest height above ground level where the wet-bulb temperature is below 0 deg C. Hail does not generally melt when the wet-bulb temperature is below zero, so when this level is closer to the ground, hail is less likely to melt as it falls, with the result being larger hailstones. WBZ heights of less than 10500 ft AGL correlate well with large hail at the surface when storms develop in an air mass primed for strong convection.

APPENDIX 9
Types of thermodynamic diagrams

Emagram

The emagram was created in 1884 by electromagnetism expert Heinrich Hertz to eliminate tedious calculations of adiabatic changes. It represents the very first chart useful for plotting atmospheric soundings. The coordinate system is linear temperature (X-axis) with quasi-linear pressure (Y-axis) and it provides straight lines for both.

Pseudoadiabatic (Stüve) diagram

Developed in 1927 by G. Stüve, the Stüve diagram is an improvement on the emagram. It is visually the simplest and most intuitive of all thermodynamic charts. It is almost identical to the emagram, however it has the added bonus of straight potential temperature lines, making three sets of perfectly straight lines. The Stüve diagram was widely used in the United States until the 1960s, when it was replaced by the skew-T.

Pastagram

Created in 1945 by J.C. Bellamy, this unusual diagram has a coordinate system of linear entropy (X-axis) and logarithmic pressure (Y-axis). It is similar to the Stüve diagram but features skewed temperature lines.

Aerogram

Created in 1935 by A. Refsdal. This was the forerunner to the skew-T diagram. Its coordinate system is linear temperature (X-axis) and logarithmic pressure (Y-axis). Equal areas represent equal amounts of work, and it was useful for determining integrated values of instability.

Tephigram

Created by Sir William Napier Shaw in 1922, the tephigram uses a coordinate system of linear reverse temperature (X-axis) and linear potential temperature (Y-axis). In the late 1940s the diagram was modified by rotating it about 45 degrees counterclockwise so that pressure lines would be horizontal, and the temperature coordinate was no longer reversed. The tephigram is structurally very similar to the skew-T but has pressure lines that curve slightly instead of being straight horizontal. This attention to detail makes it perfect for exacting physical calculations. The tephigram is still a favorite among European users, particularly in the United Kingdom.

Skew T log p

The skew T diagram, a variation of the emagram, was developed in 1947 by N. Herlofson and was quickly embraced by the U.S Air Force, gradually gaining acceptance in the National Weather Service. It's by far the most common chart used in the United States and Canada, and is what will be used throughout this book. The diagram is very similar to the modern tephigram, except that pressure lines are perfectly horizontal to make it easier to estimate altitudes.

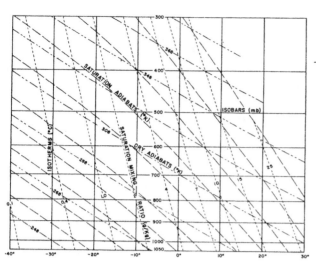

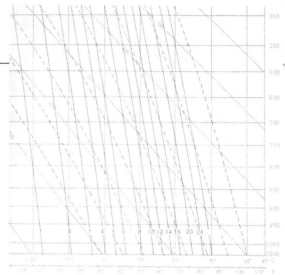

Figure 3-21a. Emagram. Developed by Heinrich Hertz in 1884, the emagram is the basis for all thermodynamic diagrams that were developed afterward. (From *The Use of the Skew-T Log-P Diagram in Analysis and Forecasting*, 1979, AWS)

Figure 3-21b. Stüve (pseudoadiabatic) diagram. This diagram was in heavy use by the Weather Bureau during much of the 1940s and 1950s. Note how the temperature and pressure coordinates form a gridlike pattern. The potential temperature line is perfectly straight (converging towards the upper left), making it easy to differentiate it from an emagram.

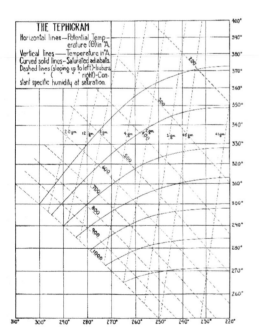

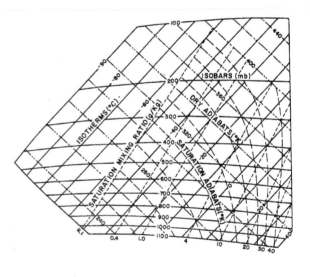

Figure 3-21c. Early tephigram. This is the classic form of the tephigram as it existed up until the late 1940s. Note how the coordinate system is linear **reversed** temperature (X-axis) with linear potential temperature (Y-axis), both labelled here in Kelvin. (From *Aeronautical Meteorology*, 1938, Taylor)

Figure 3-21d. Modern tephigram. The modern tephigram grew popular during the 1950s. This is simply the early tephigram rotated counterclockwise 45 degrees to make the pressure lines vertical. By looking at the early tephigram and turning the page in this fashion, the reader can see the change. (From *The Use of the Skew-T Log-P Diagram in Analysis and Forecasting*, 1979, AWS)

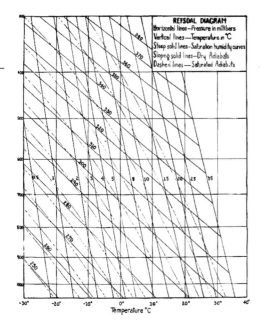

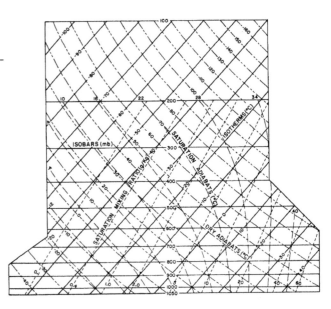

Figure 3-21e. Refsdal (aerogram) diagram. This is the forerunner to the skew-T diagram. Its coordinate system is linear temperature (X-axis) and logarithmic pressure (Y-axis). (From *Aeronautical Meteorology*, 1938, Taylor)

Figure 3-21f. Skew T log p diagram. This is the most common diagram used in the United States and Canada.

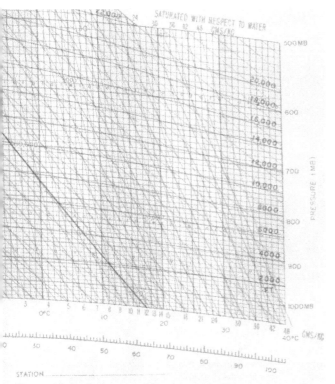

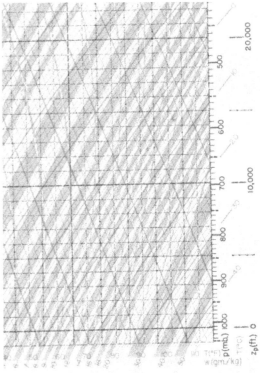

Figure 3-21g. Arowagram. This rare diagram uses a coordinate system of linear temperature (X-axis) and curved pressure (Y-axis), combining the gridlike attributes of the Stüve diagram with the logarithmic, quasi-horizontal pressure scale of the modern tephigram.

Figure 3-21h. Pastagram. This is one of the most obscure thermodynamic diagrams in weather history. It is similar to the Refsdal diagram with linear entropy (X-axis) and logarithmic pressure (Y-axis). This specimen dates to 1946.

APPENDIX 10
Blank diagrams

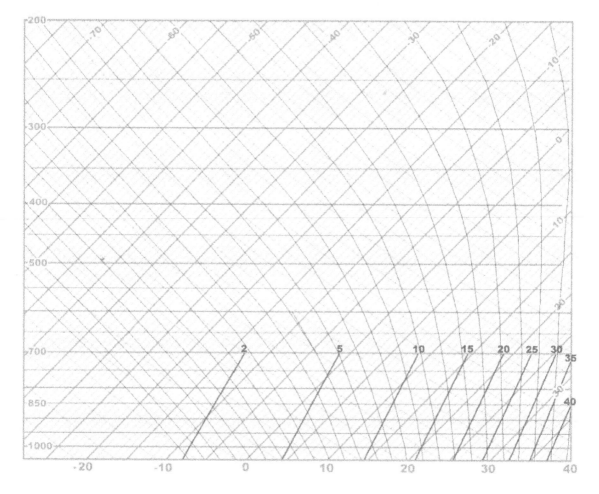

Blank Skew-T diagram. It gives us a coordinate system for plotting the vertical temperature profile through the atmosphere. Height increases toward the top of the chart, and temperature increases toward the right side of the chart. This diagram may be copied for personal use.

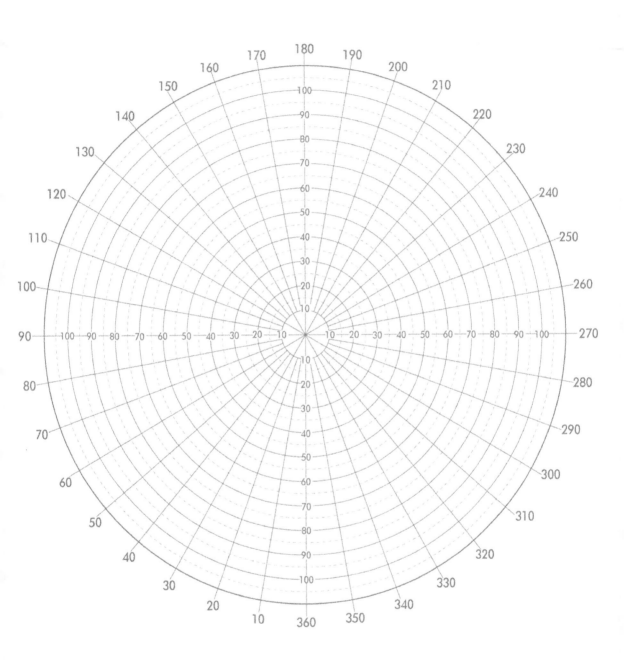

Blank hodograph. This page may be copied for personal use.

APPENDIX 11

Observation format overview

This is a quick reference guide to some of the most commonly-used weather reporting formats. For other formats, download our codes section from our web site (see the appendix).

Surface METAR format (WMO FM 15 METAR)

METAR is the most commonly-encountered format used for surface observations and is used heavily within the United States.

```
KICT 161553Z AUTO 35008G17KT 1 1/2SM -RA BR BKN011 OVC016 21/20
A3005 RMK AO2 RAB02 SLP169=
```

KICT — Station identifier for Wichita Kansas. To search or get a list of identifiers, visit the NWS's station search page on the Internet at: http://www.weather.gov/tg/siteloc.shtml

161553Z — The 16th day of the month at 1553 GMT (GMT is also known as "UTC", "Z", or "zulu" time). Eastern Time is 5 hours behind GMT time (4 hours when on daylight saving time).

AUTO — An automated station prepared the report. Some automated stations are augmented by a human.

35008G17KT — Winds are blowing from 350 deg (north) at 8 knots gusting to 17 knots (KT). Winds are always referenced to true north, not magnetic north.

1 1/2SM — Visibility is 1.5 statute miles. European stations typically report in meters or kilometers.

-RA BR — Light rain (-RA) and mist (BR) is occurring. Other symbols that may be seen are TS (thunderstorm), FG (fog), SN (snow), SH (showers), GR (hail), FZ (freezing precipitation), DZ (drizzle), and others. A minus prefix means light; a plus prefix means heavy.

BKN011 OVC016 — A broken layer (60 to 90% cover) of clouds exists at 1100 ft above ground level. A second layer of clouds, overcast (100% cover) exists at 1600 ft above ground level. A layer can be CLR or SKC (clear, 0%), FEW (less than 10%), SCT (scattered, 10-50%), BKN (broken, 60-90%), or OVC (overcast, 100%).

21/20 — Temperature and dewpoint is 21 and 20 deg C, respectively. A prefix of M means "minus".

A3005 — Altimeter setting (pressure) expressed in hundreds of inches. In this example the value is 30.05 inches of mercury.

RMK — Indicates remarks follow.

AO2 — The type of automated station is AO2 (ASOS).

RAB02 — Rain began at 2 minutes past the hour.

SLP169 — The sea-level pressure is 1016.9 millibars. If SLP is above 500 to 600, a 9 rather than 10 is prefixed, which indicates a value in the 900s range.

Surface synoptic format (WMO FM 12 SYNOP)

```
72365 11966 82504 10074 21001 39875 40157 52008 69901 70206 8807/
```

72365 — WMO Index number for Albuquerque, NM.

11966 — Coded visibility and cloud height values (will not be decoded here).

82504 — Total cloud cover in eighths is "8", winds are from 250 ("25") at 4 ("04") knots.

10074 — "1" is marker for temperature, "0" is sign (positive), and temperature is "074" (7.4 deg C).

21001 — "2" is marker for dewpoint, "1" is sign (negative), and dewpoint is "001" (-0.1 deg C).

39875 — "3" is marker for station pressure data (will not be decoded here).

40157 — "4" is marker for sea level pressure data, "0157" is pressure in mb in tens (if under 5000 add 1000 mb); in this case the pressure is 1015.7 mb.

52008 — "5" is marker for pressure change data. "2" is change type; "008" is change in hundreds of mb.

69901 — "6" is marker for rainfall data (will not be decoded here).

70206 — "7" is marker for weather occurrences; "02" (no significant weather) was occurring now; types "0" (none) and "6" (rain) had occurred during the past observation period.

8807/ — "8" is marker for cloud data. Height of lowest cloud was "8" (will not be decoded here), lowest cloud was "8" (stratocumulus), middle cloud was "7" (altocumulus), and high cloud was "/" (not observable).

Upper air radiosonde code (WMO FM 35 TEMP)

All reports start with the following information:

70026 — WMO index number: Barrow, Alaska.

TTAA/TTBB/PPBB — Indicates that mandatory level temperatures (TTAA), significant level temperatures (TTBB), or significant level winds (PPBB) follow.

66121 — The "66" is the date of the month; 50 is added when the wind speeds are in knots, otherwise winds are in meters per second. In this case, it is the 16th and winds are in knots. "12" is the GMT hour, in this case, 1200 GMT. The "1" is a code that indicates what kind of data is used in the report (not described here).

70026 — The WMO index number is repeated.

```
70026 TTAA  66121 70026 99010 05006 10006 00090 09606 10506
92745 12660 19005 85447 06636 20008 70012 02356 23011 50561
18564 20021 40723 29164 19529 30922 44959 21025 25041 55158
```

```
22524 20184 49159 21024 15375 46564 19015 10644 46165 16509
88231 57958 21031 77999 51515 10164 00005 10194 17506 22009=
```

70026 TTAA 66121 70026 — See notes above

99010 — The surface ("99") level has a pressure figure of "010" (decoding will not be provided here).

05006 — The temperature is 05.0 deg C (if the tenths digit is odd, it signifies a negative temperature). The dewpoint depression "06" indicates the dewpoint is 0.6 C degrees lower than the temperature. If the value exceeds "50" then subtract 50 to get the value in whole degrees (e.g. "58" means 8 C deg).

10006 — Winds at this level are 100 true at 06 knots.

The above 3 groups are repeated over and over.

```
70026 TTBB  66120 70026 00010 05006 11005 05807 22000 09606
33993 12001 44963 13219 55954 13256 66925 12660 77841 05823
88788 02637 99748 01108 11719 00957 22610 09538 33586 10750
44579 10960 55572 11357 66555 12760 77543 13568 88531 14960
99499 18764 11493 19556 22484 20927 33468 22924 44462 22962
55453 22567 66346 37358 77231 57958 88216 55758 99202 48959
11168 45564 22100 46165 31313 01102 81102=
```

70026 TTBB 66120 70026 — See notes above

00010 — "00" is a sequence number, which increases to 11, 22, 33, 44, etc. It's there merely to help quickly identify each group visually. The "010" is the level in millibars for the following data (if it's less than 100, add it to 1000; in this case we have 1010 mb).

05006 — The temperature at this level is 05.0 deg C and dewpoint depression is 0.6 deg C, yielding a dewpoint of 4.4 deg C. See rules above under TTAA.

The above 2 groups are repeated over and over.

```
PPBB  66120 70026 90012 10006 13504 18505 90346 19505 20007
20509 90789 22009 23008 23008 91024 23011 22016 20022 9156/
19525 20025 92035 20021 19530 20028 93035 21026 24024 21025
9447/ 19014 20012 9503/ 18009 17010=
```

PPBB 66120 70026 — See notes above.

90012 — This group indicates the height of the following data. "9" is a placeholder. "0" is the height in ten-thousands of feet. "0", "1", and "2" indicates the height in thousands of feet of the first, second, and third group respectively. So the groups to follow are for 0, 1000, and 2000 ft MSL.

10006 — Winds at 0 ft are 100 at 6 kts.

13504 — Winds at 1000 ft are 135 at 4 kts.

18505 — Winds at 2000 ft are 185 at 5 kts.

These groups repeat as necessary.

Suggested Reading

Barry, R. G. and Chorley, R. J., 1998: Atmosphere, Weather & Climate, Methuen & Co, Ltd, London. ISBN 0-415-16020-0. In print.

Djuric, D., 1994: Weather Analysis, Prentice Hall, Englewood Cliffs, NJ. ISBN 0-135-01149-3. An excellent and highly detailed primer for all enthusiasts and meteorologists. Facsimile version currently in print.

Doswell, C. A., 1982: The Operational Meteorology of Convective Weather: Volume 1, Operational Mesoanalysis, National Weather Service, Washington, D.C. Out of print but available through National Technical Information Service (www.ntis.gov) and for sale in PDF format at weathergraphics.com.

Gedzelman, Stanley David, 1980: The Science and Wonders of the Atmosphere, John Wiley & Sons, New York. ISBN 0-471-02972-6. Out of print but excellent.

Kurz, Manfred, 1990: Synoptic Meteorology, Deutscher Wetterdienst, Offenbach, Germany. ISBN 3-88148-338-1. Availability unknown.

Moore, J. T., 1992: Isentropic Analysis and Interpretation, National Weather Service Training Center, Kansas City, MO. Internal forecaster handout, not generally available except perhaps in local National Weather Service office libraries.

National Weather Service, 1993: Forecaster's Handbook #1, Washington, DC. This is mainly an overview of NWS products, not a forecasting how-to. Source of this document is the National Weather Service.

Nielsen-Gammon, J. W., 1995: Introduction to Isentropic Analysis, Texas A & M University, College Station, TX. Probably available through Texas A&M meteorology department.

Office for the Federal Coordinator of Meteorology, 2000: Federal Meteorological Handbook #1: Surface Weather Observations and Reports. Available in its entirety online at http://www.ofcm.gov/fmh-1/cover.htm . This describes the complete set of practices for taking and coding observations in the U.S.

Saucier, W. J., 1983: Principles of Meteorological Analysis, Dover Publications, New York. ISBN 0-486-65979-8. This is a reprint of an older (circa 1960s) title. Provides a complex theoretical framework for meteorological analysis.

World Meteorological Organization, 1995: Publication 306: Manual on Codes, Volume 1, Part A: Alphanumeric Codes. WMO, Geneva. ISBN 92-63-15306-X. A complete guide to the international coding formats for METAR, upper air, and dozens of other obscure types. Unfortunately it comes with a high price tag; be prepared to pay well over $100.

Software

Although the Internet itself provides a valuable source of data, most experienced forecasters realize that the Web does not provide rapid methods for analyzing, plotting, and dissecting data. Fortunately there are several powerful programs that accomplish this.

DIGITAL ATMOSPHERE. This unique software program, designed for Windows, allows flexible plotting of surface and upper charts, soundings, and climatology. It even performs 3-D roaming soundings and cross sections. Coverage is worldwide. Automatically retrieves free weather data on its own via the Internet. Digital Atmosphere can be downloaded from:

> http://www.weathergraphics.com/da/

GRLEVEL2/GRLEVEL3. For those who enjoy radar data (United States only), Mike Gibson's GRLevelX suite of radar viewers is the ideal tool:

> http://www.grlevelx.com/

RAOB. RAOB is the most extensive and detailed software available for sounding and cross section analysis. A demo of RAOB can be downloaded from:

> http://www.weathergraphics.com/raob/

Educational Websites

METEOROLOGY GUIDE. This is one of the best online weather education sites. It leans toward the basics, but there is some interesting information on weather analysis.

`http://ww2010.atmos.uiuc.edu/(Gh)/guides/mtr/home.rxml`

JOSEPH BARTLO'S ARTICLES. It's surprisingly tough to find good articles about weather analysis and forecasting on the Internet. But Joseph Bartlo has a gold mine of them. Check them out.

`http://joseph-bartlo.net/assortment.htm`

CHUCK DOSWELL'S ARTICLES. Some great articles from a leading retired National Oceanic and Atmospheric Administration researcher and an expert weather analyst:

`http://www.cimms.ou.edu/~doswell/Essays_index.html`
`http://www.cimms.ou.edu/~doswell/Web_Formalpubs.html`
`http://www.cimms.ou.edu/~doswell/inforpub.html`
`http://www.flame.org/~cdoswell/opfun.html`

ROGER EDWARDS' DATA LINKS. A great place to start to see Web-based graphics useful for forecasting purposes.

`http://www.stormeyes.org/tornado/rogersif.htm`

IWIN. All the National Weather Service bulletins you could ever want. Great for monitoring weather events.

`http://iwin.nws.noaa.gov/`

FORECAST HANDBOOK. This is the official web site for the Forecast Handbook. You can download a free section on meteorological codes. You can also download lots of weather software and look through information and products relating to weather forecasting and analysis.

`http://www.weathergraphics.com/`

THE FAQ FOR HURRICANES, TYPHOONS, AND TROPICAL CYCLONES. This resource by Christopher W. Landsea is the definitive starting point for finding out more about tropical storms.

`http://www.aoml.noaa.gov/hrd/tcfaq/tcfaqHED.html`

Government weather agency websites

Following are web site addresses of official government weather agencies. If you get a "404 Error" or the link no longer works, you can often try shortening the URL by removing data that follows the "/" characters. For translation to English from any of several major languages, try the Google Translator at http://translate.google.com/ . For other sites visit the WMO link below.

Argentina	`http://www.smn.gov.ar/`
Australia	`http://www.bom.gov.au/`
Brazil	`http://www.inmet.gov.br/`
Canada	`http://www.msc.ec.gc.ca/`
China	`http://www.cma.gov.cn/`
Cuba	`http://www.met.inf.cu/`
France	`http://www.meteo.fr/`
Germany	`http://www.dwd.de/`
Italy	`http://www.meteoam.it/`
Japan	`http://www.jma.go.jp/`
Korea	`http://www.kma.go.kr/`
Kenya	`http://www.meteo.go.ke/`
Mexico	`http://smn.cna.gob.mx/`
Netherlands	`http://www.knmi.nl/`
NZ	`http://www.metservice.com/`
Norway	`http://www.met.no/`
Pakistan	`http://www.pakmet.com.pk/`
Philippines	`http://www.pagasa.dost.gov.ph/`
Poland	`http://www.imgw.pl/`
Russia	`http://www.meteorf.ru/`
Singapore	`http://www.nea.gov.sg/metsin/`
Saudi Arabia	`http://www.pme.gov.sa/`
South Africa	`http://www.weathersa.co.za/`
Spain	`http://www.aemet.es/es/nuevaweb/`
Sweden	`http://www.smhi.se/`
Switzerland	`http://www.meteoswiss.ch/`
Taiwan	`http://www.cwb.gov.tw/`
Thailand	`http://www.tmd.go.th/`
Turkey	`http://www.meteor.gov.tr/`
UK	`http://www.metoffice.gov.uk/`
USA	`http://www.weather.gov/`
WMO	`http://www.wmo.int/`

Ten weather myths

10. "MAMMATUS INDICATES TORNADOES."
It's often seen on the weakest of storms, such as those in the tropics, and even on the underside of altostratus layers.

9. "UPPER COLD ADVECTION RESULTS IN DESTABILIZATION." Air follows isentropic surfaces, not constant pressure surfaces. Therefore a pattern of cold advection as seen on a constant-pressure chart (such as 500 mb) does not necessarily mean cooling will occur.

8. "TORNADO INTENSITY IS MEASURED WITH THE FUJITA SCALE." The Fujita scale is a damage scale, not an intensity scale.

7. "UPPER Diffluence MEANS UPWARD MOTION." Diffluence is frequently cancelled out by speed convergence.

6. "FALLING SNOW WILL MELT WHEN IT'S OVER 32°F." If the air is dry, any melting actually causes heat to be removed from the air, resulting in cooling. The wet bulb temperature must be above freezing for the snow to melt.

5. "POSITIVE VORTICITY ADVECTION MEANS UPWARD MOTION." Seeing PVA at 500 mb only provides an inference of one term of the omega equation for vertical motion. It is frequently cancelled out by thermal advection.

4. "TROPICAL CYCLONES DIE AFTER MAKING LANDFALL BECAUSE OF FRICTION." Research has shown that factors such as the lack of moisture and heat sources are the real catalyst for dissipation. Ironically, the friction just after landfall actually increases turbulent flow and brings stronger winds down to the surface.

3. "HIGHER RESOLUTION MODELS MEAN BETTER FORECASTS." Operational models as fine as several kilometers are now in use. Unfortunately the models are only as good as the data that feeds them. We are still sampling the atmosphere using an average upper-air spacing of at least 500 km and a surface spacing of 100 km, though arguably the improvements in the surface network plus the integration of remote sensing data like radar may be the primary factors that have resulted in better model success.

2. "THAT TORNADO OCCURRED BECAUSE OF EL NINO / LA NINA / GLOBAL WARMING, ETC." A climatological phenomena with a cycle of six months or more cannot possibly cause a ten-minute weather occurrence. There are indications that variations in the large-scale circulation may have some influence on the environment to favor severe thunderstorms in one area of the country over a long-term period, but in the end tornadoes are still caused by small-scale processes in the thunderstorm.

1. "TOILETS SPIN THE OTHER WAY IN THE SOUTHERN HEMISPHERE." Maybe this isn't a forecasting myth, but forecasters frequently have to explain the Coriolis force. It's no doubt that tourist traps on the Equator combined with questionable travel shows on television have done quite a bit to perpetuate this myth. Simple equations show that the contribution of the Coriolis force to such a small-scale process is infinitesimally small, and is vastly overpowered by the trajectory of water entering the bowl as conservation of angular momentum imparts spin to the fluid. In other words, you can pour a bucket of water into the toilet and control the spin in either direction. For more information, see: http://www.ems.psu.edu/~fraser/Bad/BadCoriolis.html

Index

Symbols

20R85 rule 166
30R75 rule 166

A

absolute instability 39
absolute vorticity 13
accretion 120
adiabat
  dry 42
  wet 42
advection 68
  differential 68
advection jet 78
advection lobe 77
aerogram 225
air mass 87
aliasing
  velocity 151
altocumulus 204
altostratus 208
anafront. *See* front: active
analysis 183
  numerical 190
angular velocity 8
anticyclogenesis
  in arctic air masses 118
anvil 202
AO 196
arctic
  air mass 90, 116
attenuation
  of satellite imagery 131
AVN. *See* Spectral model

B

baroclinic
  high 108
  low 103
baroclinic instability 103
baroclinic leaf 137
baroclinic zone cloud system
      137
barotropic

high, cold-core 114
high, warm-core 115
low, cold-core 110
low, warm-core 112
Bayesian averaging 191
block numbers 23
bomb 105
boundary errors 194
bounded weak echo region
      153
bow echo 163
box technique
  thickness 71
braking 109
BRN 223
BRN shear 223
BWER 153

C

CAA. *See* advection: thermal
cap 44
CAPE 223
capping inversion 44
chimney effect 60
CIN 223
CINH 223
cirrocumulus 205
cirrostratus 207
cirrus 206
classic supercell 160
closed-cell
  cloud layers 135
clouds
  and satellite imagery 132
  observation 28
cold air advection 71
cold conveyor belt 106
cold front 93. *See* front:
      cold
cold occlusion
      95, 96. *See* occlusion:
      cold
cold sector 92
comma cloud 138
conditional instability 39
constant pressure chart 54
contamination
  of satellite imagery 131
continental polar 88
continental tropical 89

contours 14
convective feedback 194
conveyor belt 105
coordinate systems 13
Coriolis 8
Coriolis force 18
correlation coefficient 147
cross section 82
cross totals index 224
CT 224
cumulonimbus 202
cumulus 201
cutoff high 115
cutoff low 112
cyclostrophic wind 11

D

damper effect 61
dealiasing 151
deep easterlies 172
deformation zone
  and frontogenesis 72
dendrite 121
dendritic growth zone 120
density 5
  and thickness 70
derecho 164
deviant motion 166
dewpoint depression 7
dewpoint temperature 7
diabatic processes
  and isentropic analysis 80
  and winter weather 123
diagnosis 185
differential advection 68
differential phase shift 147
differential reflectivity 147
diffluence 59
direction 2
divergence
  frictional 62
downdraft 157
dry adiabat 42
dry conveyor belt 106
dryline 96, 98
  location 96
  movement 97
  structure 96
dry snow 124
dual-polarization radar 146

dust 97, 140
dynamical models 189

E

easterly wave 172
EHI 223
elevated front 107
elevated mixed layer (EML)
 44, 96
El Niño 195
Elvis' UFO 13
emagram 225
EML 44, 96
energy-helicity index 223
ensemble forecasting 191
equatorial
 air mass 90
equatorial trough 169
equivalent potential temperature 50
explosive cyclogenesis 105
extratropical cyclone 103
eyewall 176

F

ferrel cell 18
forecasting 183
foreshortening 130
free atmosphere 17
freezing rain 120
friction 10, 61, 62
front
 active 93
 anafront. *See* front: active
 and satellite imagery 135
 elevated 107
 inactive 93
 inversion 91
 katafront. *See* front: inactive
 location 90
 movement 91
 slope 91
 split 107
 stationary. *See* front:
  quasistationary
 surface 91
frontal inversion 43, 91
frontal low 103

frontogenesis 71, 92
 and q vectors 78
 forcing 73
 response 73
frontolysis 71, 92
 and q vectors 78

G

geostationary satellite 127
geostrophic wind 9
GFS model 192

H

hadley cell 17
Hadley cell 171
heat low 113
Henry's rule 112
heterogenous nucleation 120
high-precipitation supercell
 160
hodograph 229
hurricane 113. *See* tropical
 cyclone
Hydrometeor Classification
 Algorithm 148

I

ice pellets 120
ID method 166
inflow 168
infrared imagery 129
initialization 190
instability 157
Intertropical Convergence
 Zone 170
inversion 43
 capping 44
 frontal 43
 radiational 43
 subsidence 44
isentropic lift 81
isentropic surfaces 79
isobar 86
isodrosotherm 87
isohypse 54
isopach 70
isotherm 86
isothermal vorticity advection
 75, 78

ITCZ 170

J

jet 62
jet max. *See* jet streak
jet streak 66
 quadrants 67
jet stream. *See* polar front jet
 and satellite imagery 139

K

katafront. *See* front: inactive
Kelvin 4
KI 224
K-Index 224
Kona low 174

L

land breeze 99
La Niña 195
latent heat 36, 37, 113
latitude 8
lee cirrus 140
level of nondivergence 60
LEWP 163
LI 224
lift 157
Lifted Index 224
line echo wave pattern 163
LLJ 64. *See* low level jet
location identifiers 23
long wave 54
 and satellite imagery 136
 polar vortex. *See* polar
  vortex
low level jet 64
low-precipitation supercell
 160

M

maritime polar 89
maritime tropical 90
mass 3
math 1
MCC 164
MCS 162
mesoscale 15

mesoscale convective system 162
METAR 24, 231
microscale 15
mid-tropospheric cyclone 173
mixed layer (ML) 45
mixing ratio 6
  and SKEW-T's 40
moist adiabatic lapse rate 37
moisture 157
momentum 48, 50, 97
monsoonal trough 171
most unstable (MU) 46
multicell
  cluster 158
  line 159
  MCC. *See* MCC

N

NAM model 192
NAO 196
NEXRAD 143, 153
nimbostratus 209
non-divergence
  level of. *See* level of non-divergence
numerical models 188
NVA. *See* vorticity: advection

O

observations 21
occluded front
  95. *See* front: occluded
  cold occlusion 95
  warm occlusion 96
occlusion 104, 110
okta 29
omega equation 74
open-cell
  cloud layers 135
open wave 104

P

parcel
  lifting 45
pastagram 225
PBL. *See* planetary boundary layer
PFJ 62. *See* polar front jet
phase, of matter 35
phase shift
  differential 147
planetary boundary layer 16
plateau high 116
PNA 195
polar cell 19
polar front jet 62
polar low 107
polar-orbiting satellite 127
potential instability 48
pressure 26, 27
Pressure coordinate system 13
pressure gradient 8
pressure surface 14
prevailing visibility 27
prognosis 186
pseudo-adiabatic 37
pseudoadiabatic diagram 225

Q

QFE 26
QLM. *See* Quasi-Lagrangian model
QNH 27
quadrants
  divergent/convergent 67
quasistationary front
    94. *See* fronts: quasista-
    tionary
Q-vector 78

R

radar 143
radiation
  and arctic air masses 117
radiational inversion 43
radiosonde 30
  code format 232
RAFS. *See* NGM
range folding 150
rear inflow jet 163
rear inflow notch 163
reflectivity 143
  differential 147
reflectivity gradient 152
relative vorticity 12
right hand rule 12
riming 120
Rossby number 11
Rossby wave 54
RUC model 193

S

SAO 25
satellite 127
  infrared imagery 129
  visible imagery 129
  water vapor imagery 130
saturation mixing ratio 6
scale 2, 3, 15
  mesoscale 15
  microscale 15
  planetary 15
sea breeze 99
sectors 92
SFD. *See* forecast discussion
shallow easterlies 172
shear lobe 77
short wave 58
  and satellite imagery 137
Showalter Index 224
SI 224
skew T log p diagram 39, 225, 228
slantwise convection 50
sleet. *See* ice pellets
sloshing 98
snow 120
  dry 124
  wet 124
specific differential phase 148
spectrum width 146
speed divergence 60
split 168
split front 107
squall line 162
SRH 223
stability index 47
stationary front. *See* fronts: quasistationary
station plot
  surface 211
  upper air 213
station pressure 26
STJ 63. *See* subtropical jet

storm motion 166
storm movement 166
storm-relative helicity
    168, 223
stratocumulus 203
stratosphere 16
stratus 210
Stüve diagram 225
subgeostrophic wind 66
subsidence 37
  and baroclinic highs 109
subsidence inversion 44
subtropical high 115
subtropical jet 63
subtropical ridge 171
supercell 160
  classic 160
  high-precipitation 160
  low-precipitation 160
supercooled water 120
supergeostrophic wind 67
Superior air mass 44
surface aviation observation
    25
surface based (SB) 45
surface chart 85
SWEAT 224
symmetric instability 48
SYNOP 24, 232
synoptic observations 24
synoptic scale 15

T

Tau technique 124
tephigram 225
thermal gradient 103
thermal low 113
thickness 69
thunderstorm 157
  motion of 166
total totals index 224
trade wind 172
transverse band 134
transverse circulation 68
triple point 104
tropical cyclone 113, 174
  forecasting 178
tropical upper tropospheric
    troughs 174
troposphere 16

trough 96, 97
TT 224
TUTT 174
TUTT low 174
typhoon 113. *See* tropical
    cyclone

U

updraft 157
upper air chart
  analysis of 214

V

VAD 154
vector 3
velocity 2
  radar 144
velocity azimuth display 153
velocity wind profile 153
ventilation 168
  of storm 168
vertical motion 60
vertical totals index 224
virtual temperature 7
visible imagery 129
vorticity 11
  advection lobe 77
  and jet streaks 75
  and vertical motion 73
  isothermal advection 75
  isothermal advection of 78
  lobe 77
  shear lobe 77
VT 224
VWP 153

W

WAA. *See* advection: ther-
    mal
warm air advection 71
warm conveyor belt 106
warm front 93. *See* front:
    warm
warm occlusion 96. *See* oc-
    clusion: warm
warm sector 92
water vapor imagery 130
WBZ 224
weak echo region 153

WER 153
wet adiabat 42
wet adiabatic lapse rate 37
wet bulbing 123
wet bulb temperature 7
wet-bulb thermometer 25
wet-bulb zero 224
wet snow 124
wind vectors 166
winter weather 119
WRF model 192

CPSIA information can be obtained
at www.ICGtesting.com
Printed in the USA
LVHW022017200623
750260LV00006B/696

9 780983 253303